Blohm & Voss Bv 141

David Myhra

Schiffer Military History
Atglen, PA

Acknowledgments

- *Gary Webster* - thanks for all your help digging out many photographs, graphs, and diagrams found in this illustrated history of the asymmetrical *Blohm & Voss Bv 141.* Your archival resource fullness is nothing short of amazing, and I do believe, given time, you could even locate a *Bv 141* were one hiding somewhere out there.

- *Reinhard Roeser* - fine-scale model kit builder, none better. Photos of his scale model *Blohm & Voss Bv 141* appear throughout this illustrated history.

- *Ronnie Olsthoorn* - creator of the very beautiful digital images of the *Blohm & Voss Bv 141* found within.

Book Design by Ian Robertson.

Copyright © 2001 by David Myhra.
Library of Congress Control Number: 2001092665

All rights reserved. No part of this work may be reproduced or used in any forms or by any means – graphic, electronic or mechanical, including photocopying or information storage and retrieval systems – without written permission from the copyright holder.

"Schiffer," "Schiffer Publishing Ltd. & Design," and the "Design of pen and ink well" are registered trademarks of Schiffer Publishing, Ltd.

Printed in China.
ISBN: 0-7643-1397-5

We are interested in hearing from authors with book ideas on military topics.

Published by Schiffer Publishing Ltd.
4880 Lower Valley Road
Atglen, PA 19310
Phone: (610) 593-1777
FAX: (610) 593-2002
E-mail: Schifferbk@aol.com.
Visit our web site at: www.schifferbooks.com
Please write for a free catalog.
This book may be purchased from the publisher.
Please include $3.95 postage.
Try your bookstore first.

In Europe, Schiffer books are distributed by:
Bushwood Books
6 Marksbury Avenue
Kew Gardens
Surrey TW9 4JF
England
Phone: 44 (0) 20 8392-8585
FAX: 44 (0) 20 8392-9876
E-mail: Bushwd@aol.com.
Free postage in the UK. Europe: air mail at cost.
Try your bookstore first.

Blohm & Voss Bv 141

What you are about to read regarding the birth and death of the *Blohm & Voss Bv141*—the prototype asymmetrical short-range reconnaissance and army ground support aircraft of 1938—is pretty much at odds from what has been presented in books on German aircraft of *WW2.* This machine's planform was different, very different from conventional aircraft designs then or since. Consequently, it has been labeled an oddity due to its asymmetrical design, hence, because it was odd, fought a losing battle to get *RLM* acceptance and financial support. Not exactly. This machine would have become part of the *Luftwaffe* stable of aircraft had it not been for the fact that it required the powerful 1,560 takeoff horsepower (up one minute's operation) *BMW 801A 14*-cylinder, radial, air-cooled, piston engine. *Dr.-Ing. Richard Vogt, Blohm & Voss Flugzeugbau's* gifted director and chief of design, had been planning to install a *BMW 801A* in his *Bv 141* as early as January 1939. Instead, he had to settle for the 865 takeoff horsepower (up to one minute's operation) *BMW 132N* 9-cylinder, radial, air-cooled engine, which the prototype (*Ha 141-0*) was equipped. Actually, the *132N* was rated at 665 horsepower at 14,760 feet [4,500 meters]. At the time it appeared that *Blohm & Voss* would never have the *BMW 801A* for series production, because *BMW Flugmotorenwerke GmbH* was unable to make their production quota even for the *Dornier Do 217* twin-engine bomber, and later the *Focke-Wulf Fw 190* fighter. This is the reason why the *Bv 141* was canceled after about twenty machines were constructed...not because the *OKL* or *Luftwaffe* High Command considered it an oddity, unorthodox , and visually unappealing due to its asymmetrical planform and, therefore, plain wrong.

Richard Vogt said he recalled that the asymmetrical planform came about as a result of a new request for proposals issued in early 1937 by *General Ernst Udet*, director of the Development Section within the *Technischen Amt*, or Technical Office of the *RLM. Udet* had occupied this planning position since January 1936. *Arado Flugzeugwerke's* single engine *Ar 198*, along with *Focke-Wolf* with its *Fw 189* were *Vogt's* two other competitors. Thinking about what *Udet* needed, not just what he wanted, gave *Dr. Vogt* the impetus to dream up a most unconventional and daring design for a single-engine short-range reconnaissance aircraft...a design which proved to be very successful, as well. The request for proposals from the

Blohm & Voss' response to the RLM's call in early 1937 for a single-engined, three-seat "special purpose" machine capable of short range reconnaissance and army ground support duties...their asymmetrical Bv 141B. Scale mode by Reinhard Roeser.

RLM implied a single-engine layout, but they did not say the machine could only have a single engine...it could have twins, cockpit cabin with room for a crew of 3, and a large, wide field of view forward and downward. Furthermore, the *RLM's* specifications also stipulated good cannon protection at its rear. In addition, this machine should also be capable of fulfilling the light bomber, low-level attack, and smokescreen-laying roles for the *Wehrmacht* in an emergency. *Vogt* later said that when he began sketching up such an airplane, with the required 25% view down over the engine that it led to a pitifully large fuselage. What *Udet* (*RLM*) really needed, *Vogt* reasoned to himself and his design staff, is a two-engine airplane where pilot and observer would sit up front with a completely unobstructed view. So, why not a two-engined airplane with one of the engines removed?

Reflecting on such an asymmetrical contraption, *Vogt* recalled that it struck his mind that a symmetrically built single-engine airplane does not at all behave symmetrically in flight. The rotating slipstream behind the propeller hits the vertical tail fin (rudder) from the side, and thus constantly tends to force the airplane off its course. This, then, has to be compensated by mounting the fin to the fuselage at an angle. Could it be, *Vogt* continued to reason, that the turning-off tendency due to the rotating slipstream might be compensated by the propeller thrust of an engine, displaced sidewards, off the center of the airplane? *Vogt* said that a quick numerical analysis showed that the offset moment of the propeller thrust would be of the same order of magnitude as that of the rotating slipstream hitting the fin from the side. So, with his asymmetrical ideas in mind, *Vogt* said that he drew up the airplane and went to Berlin, directly to *Ernst Udet*. *Vogt* knew him, and he found *Udet* to be very surprised by the asymmetrical design, but not unaware of its theory or history. *Udet*, a fighter pilot in the German *Luftwaffe* from *WWI*, was the fourth highest scoring ace in the world from that conflict. *Vogt* recalled how *Udet* spoke about the giant *Gothaer Waggonfabrik*-built "*Gotha G*" bombers, which had terrorized England, and how *Gothaer* had obtained several patents for an asymmetrical bomber aircraft, too. *Gothaer's* purpose at that time with the asymmetrical planform was to obtain better fields of cannon fire and better viewing to more accurately drop their bomb load. *Vogt* recalled post war in the United States how he and *Udet* discussed the theory regarding asymmetrical planforms. In conventional symmetrical piston-powered aircraft its propeller gave the air mass it has sucked in a spinning motion, rotating it back against and around the vertical rudder of the aircraft. Throughout flight, especially during takeoff, the pilot had to constantly fight the tendency of the aircraft to brake off to port, so the pilot had to continually fight or reject the spin effect of propeller thrust upon the rudder. *Udet* gave the order allowing *Vogt* to construct three experimental prototypes, each with different fuselage lengths and wing spans, however, all with their crew cabin shifted laterally to starboard (needed in this case due to the counterclockwise rotation of the engine/propeller), adding:

The *Blohm & Voss Bv 141* would also be required to do light bombing duties (4x*SC 50* kilogram bombs), low level strafing, and smoke-laying roles for the *Wehrmacht* (army). Therefore, a need for outstanding all-around vision of the pilot and crew. Scale model and photographed by *Reinhard Roeser*.

A pen and ink illustration of the *Blohm & Voss Bv 141B* and powered by a single 14-cylinder *BMW 801N* radial engine capable of 1,560 takeoff horsepower for up to one minute's operation. Courtesy: *Hermann Pohlmann, Chronik Eines Fluggeugwerkes, Motorbuch Verlag, Stuttgart, West Germany.*

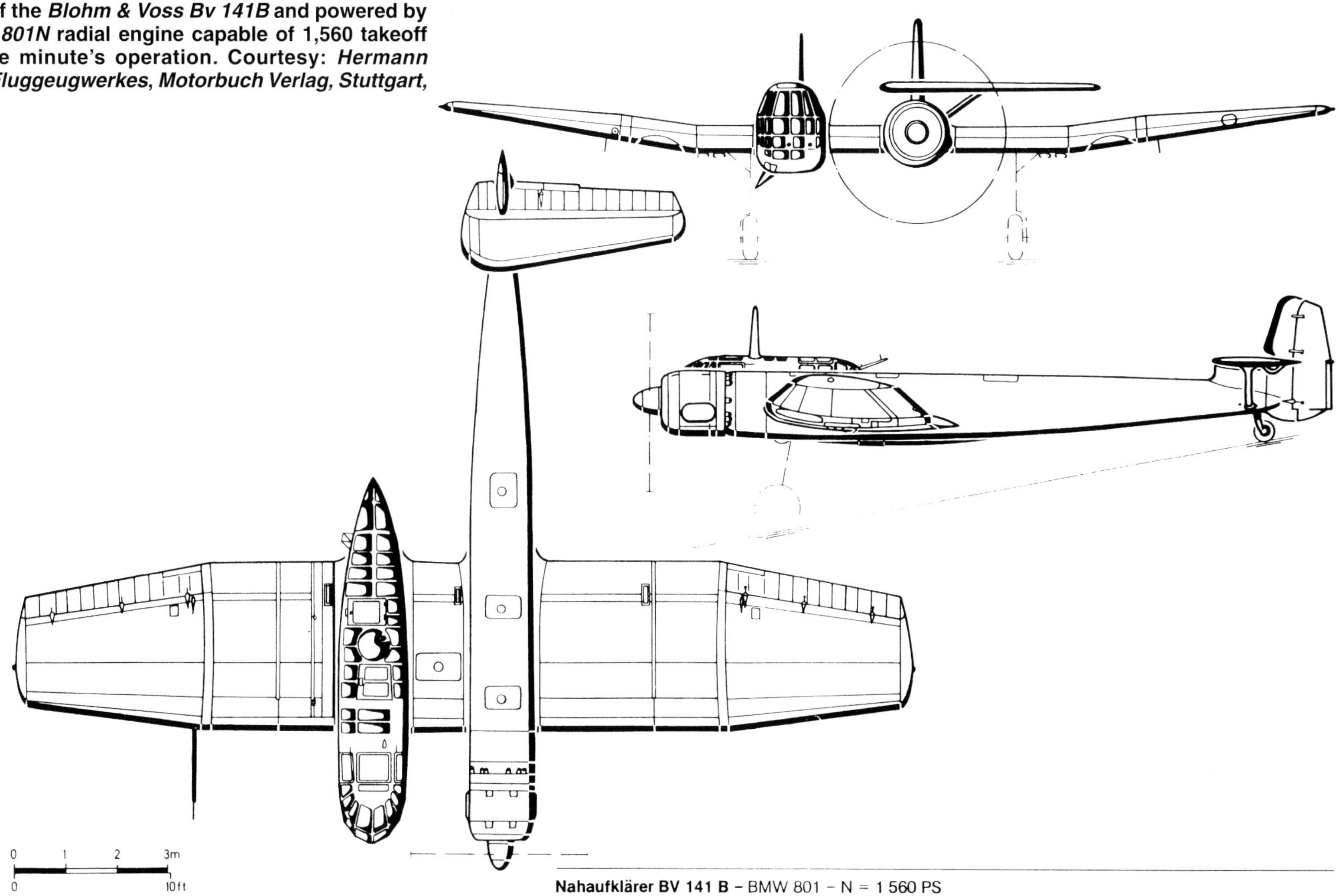

Nahaufklärer BV 141 B – BMW 801 – N = 1 560 PS

Abmessungen		Gewichte		Leistungen	
Spannweite	17,46 m	Leergewicht	4 645 kg	Höchstgeschwindigkeit	435 km/h
Größte Länge	13,95 m	Rüstgewicht	4 735 kg	Gipfelhöhe	10 000 m
Flügelfläche	53,10 m^2	Besatzung	300 kg	Reichweite	1 200/1 900 km
		Kraftstoff	800 kg		
		Nutzlast	265 kg		
		Fluggewicht max.	6 100 kg		

Flächenbelastung: 115,0 kg/m^2

Leistungsbelastung: 3,9 kg/PS

We have so much control trouble with ordinary airplanes, so I do not think that you will succeed in a year's time to be ready for the competition with your highly unusual design.

Vogt believed that his preliminary calculations showed so much promise, and that *Udet* was so excited about the possibilities that he saw no good reason why he would be unable to compete with the other two conventional types of airplanes coming from *Arado* and *Focke-Wulf.* Back at Hamburg, *Vogt* assembled a group of his best engineers, and he said that he spent most of his time at their drawing boards, hunting for really good design solutions. When the year was over his airplane was ready to fly and to compete.

Richard Vogt recalled that the first flight of the *Bv 141* was really exciting. His heartbeat quickened when the *Blohm & Voss* company test pilot lifted off this airplane (*Hamburger flugzeugbau Ha 141-0* registered *D-ORJE*) with so many unknowns on February 25th, 1938. To everyone's surprise, its initial test flight went not only well, but perfectly. *Vogt* said that he telephoned *Udet* and he came to Hamburg. Taking the *Hamburger Ha 141-0* prototype up by himself, *Udet* made loops and rolls. *Dr. Vogt's* placement of the cockpit gondola, which accommodated the displaced center of gravity which now was located to the outside of the engine's long fuselage, the pilot's nacelle perfectly counteracted the 865 take-off horsepower *BMW 132N's* propeller torque during take off. Landing, *Udet* was excited. He had the prototype refueled and flew it to Berlin, where he showed the unusual airplane to *Generalfeldmarschall Erhard Milch* (*Hermann Göring's* deputy at the time) and *Reichsminister der Luftwaffe Hermann Göring* himself, the Commander in Chief of the *German Luftwaffe.*

Udet's appraisal was well founded. A few months later *Blohm & Voss* appeared to have won the competition over the conventional (symmetrical planform) *Arado Ar 198* and *Focke-Wulf Fw 189* airplanes. *Vogt* recalled postwar that he had received an order from *Udet* for 500 of the machines. But it was a pity that *Vogt's* exciting story was to have a poor ending. About 20 machines had been already delivered (according to *Vogt's* postwar public comments in the United States), but then the whole project was suddenly canceled by the *Luftwaffe's OKL*, or *Luftwaffe* High Command. This action took place on April 4th, 1940, in favor of the *Fw 189.* Why? Because the *OKL* did not like the way the *Bv 141* looked? This myth persists today, however, the main reason was due to the *BMW 801A* radial engine. The *Bv 141* had more range and endurance than the *Fw 189.* The "double-star" *BMW 801A* engine was the same as the one used in the *Dornier Do 217* twin-engine bomber, and later in the *Focke-Wulf Fw 190* day fighter. Since *BMW* was delivering only about half as many as were needed and had promised, *Udet* said that he had no other choice:

We can live without a new reconnaissance plane, but we cannot live with a drastically reduced number of fighters and bombers.

Another factor which contributed to the *Fw 189's* success over the *Bv 141* was to ensure a continued supply of the *Fw 200 Condor. Focke-Wulf* had lost factory space where the *"Condor"* was being assembled due to bomb damage. The *RLM* gave *Focke-Wulf* 80% of *Blohm & Voss'* factory space to assemble *Fw 200s*...space which would have been used to manufacture *Bv 141s.* So, it appears that the *Bv 141* cancellation was more due to being a victim of circumstances than because it was considered ugly. Nonetheless, *Dr. Goebbels,* the *Reich's* propaganda minister, put out a great deal of propaganda about the

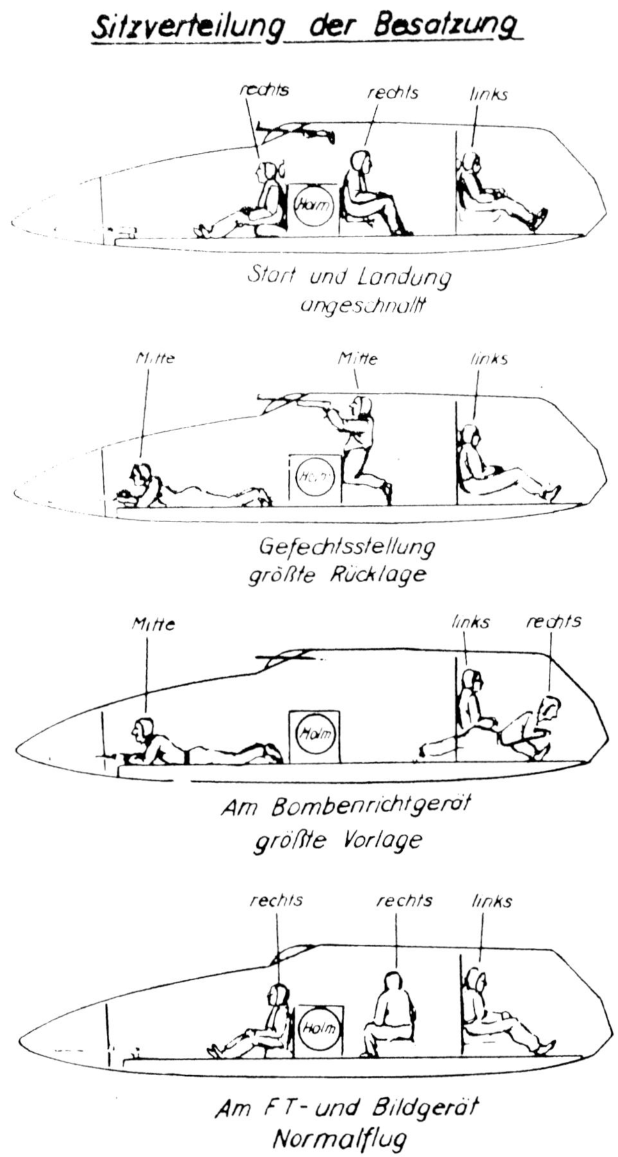

Right: A pen and ink illustration featuring the several seating arrangements in the typical *Bv 141* for its three-man crew as seen from its starboard side.

01 - seating placement during takeoff and landing;
02 - seating during in flight attack from the rear;
03 - seating during army ground support activities... bombing and/or strafing;
04 - seating during normal flight;

Bv 141 fighting on the Eastern Front in the Summer of 1942. But the asymmetrical *Bv 141* never did get to experience combat operations...on the Eastern Front or any other, so it will never be known how well it would have performed in the field compared to the *Fw 189*, even though the *RLM* ordered a squadron of *Bv 141s* to be formed and known as *"Special Squadron-Bv 141."* Due in part to the machine's teething troubles and difficulty in training aircrews, *"Special Squadron-Bv 141"* did not come to be, and the whole idea was canceled by the *RLM* in Spring 1942. It appears, however, that no one bothered to inform *Dr. Goebbels*, because he was still telling the world about its great success on the Eastern Front well into Summer 1942.

What did the world's aviation community think of *Vogt's* new machine? Well, the world's only asymmetrical airplane was later sketched and described in the "*Oddentifications*" section of the British aviation magazine "The *Aeroplane*:"

Your nose is here, your engine's there.
Your tail's just stuck on anywhere.
You symbolize the grotesque Hun,
Begone, distorted one four one!

Blohm & Voss Bv 141 Prototype Construction Sequence

It is not exactly known for certain just how many of *Richard Vogt's* asymmetrical *Blohm & Voss Bv 141s* were built before *OKL* awarded a contract for a new short range reconnaissance and *Wehrmacht* support aircraft to *Focke-Wulf*. Several aviation historians put the number at about 13 machines. *Vogt* claimed post war that about 20 machines were built. This author will accept *Richard Vogt's* number of twenty units, because thirteen seems too few given the photographic evidence. There are photos of *Bv 141* machines with radio call codes which do not fit into identifiable categories, meaning that there are six or seven so-called *Bv 141* "wild cards" which were constructed, thus bringing the total number of machines constructed to about twenty.

Hamburger Ha 141-0 - The first prototype. Construction started in March 1937, registered as *D-ORJE*, and made its maiden flight on February 25th, 1938. This is the machine which *Ernst Udet* flight tested doing loops and rolls at Hamburg, and after refueling took it to Berlin to show *Erhard Milch* and *Hermann Göring*. Relatively trouble free during its initial flight testing. Redesignated *Bv 141A V2* to reflect the company's name change to *Blohm & Voss Flugzeugbau* from *Hamburger Flugzeugbau*. The *Bv 141A V2's* initial stepped-crew nacelle was entirely redesigned (the *RLM* requested this change) in the second prototype constructed, which came to be known as *Bv 141A V1 werk numer 171*, and after its redesign the cockpit cabin came to look surprisingly similar to that found

A *Blohm & Voss Bv 141B* as seen beneath and its port side. Scale model by *Reinhard Roeser*.

on *Focke-Wulf's* twin *Argus As 410* engine-powered *Fw 189.* The *Hamburger Ha 141-0* became one of the three experimental prototypes ordered by *Ernst Udet.* Final disposition unknown.

Specifications:
- Wing span - 49 feet 2 inches [15.00 meters]
- Wing area - 446.7 square feet [41.50 meters squared]
- Length, overall - 36 feet 4 inches [11.10 meters]
- Height, overall - 10 feet 8 inches [3.30 meters]

Blohm & Voss Bv 141A V1 - *Werk Numer 171*, registered as *D-OTTO.* Second of the three experimental prototypes ordered by *Ernst Udet.* Flight trials began in September 1938. A major problem surfaced immediately involving the landing gear's hydraulic system. Sporadically it failed to function, and the cause seemed to defy attempts to fix it. The *Bv 141A V1* weighed 6,817 pounds empty and 8,441 pounds fully loaded. The three man crew's nacelle was changed with a great many additional plexiglass panels and taking on the appearance of the cockpit nacelle of the *Fw 189.* The *Bv 141A V1's* crew nacelle measured 3 feet 11 1/4 inches side to side and 4 feet 11 inches from top to bottom. Like the crew nacelle on the *Fw 189*, the *141A V1's* nacelle ended in a pointed cone. No offensive or defensive armament was installed, however, provisions had been made to have the *Bv 141A V1* include 2x*MG* 7.9 mm fixed forward-firing cannon and 2x*MG* 7.9 mm hand-held aft-firing cannon. Armament would not be installed until the *Bv 141A V3* version. In addition, the aircraft contained racks on the undersurface of the wing for carrying 4x*SC 50* kilogram [110 pound] bombs. The overall size of the *Bv 141A V1* was enlarged some beyond the initial *Hamburger Ha 141-0* (*Bv 141A V2*) machine. Its wing span was increased from 49 feet 2 1/2 inches to 49 feet 6 1/2 inches. Its wing area went from 446.7 square feet to 449.9 square feet in the *Bv 141A V2*, and its length overall went from 36 feet 5 inches to 37 feet 4 3/4 inches. Its horizontal tailplane was almost symmetrical, with trim tabs installed in each elevator. A 106 gallon fuel tank was mounted aft of the engine firewall. All control surfaces were activated by rods, except the flaps, which were hydraulically operated.

The *Bv 141A V1* was being actively tested, however, it was lost to further flight testing when it was forced to make a belly landing in a ploughed farm field on October 5th, 1938, with the main wheels only partially extended. Considerable damage was done to the *Bv 141A V1's* undercarriage and starboard wing. It is not known to this author if the *Bv 141A V1* was repaired and made flight worthy again. However, this machine was declared by *Blohm & Voss* to be their "official" prototype, even though the *Ha 141-0* (*Bv 141A V2*) was the first machine of this type to be constructed at Hamburg. Final disposition unknown.

Specifications:

- Wing span - 49 feet 5 inches [15.10 meters]
- Wing area - 450 square feet [41.8 meters squared]
- Length, overall - 36 feet 7 inches [11.40 meters]
- Height, overall - 10 feet 5 inches [3.20 meters]
- Weight, take-off - 8,441 pounds [3,829 kilograms]
- Weight, empty - 6,812 pounds [3,090 kilograms]
- Speed, at 11,482 feet [3,500 meters] - 228.7 mph [368 km/h]
- Weight, fuselage - 2,325.8 pounds [1,055 kilograms]
- Weight, engine - 1,731 pounds [785 kilograms]
- Weight, fuel - 727.5 pounds [330 kilograms]
- Weight, 3 man crew - 441 pounds [200 kilograms]

A nose port side view of a *Blohm & Voss Bv 141B*. Scale model and photographed by *Reinhard Roeser*.

Bv 141A V2 - *Werk Numer 172,* registered as *D-ORJE* and radio call code *PC+BA*. This machine was the redesigned and redesignated the *Hamburger Ha 141-0* prototype after *Blohm & Voss's* aircraft division's name of *Hamburger* was changed to *Blohm & Voss Flugzeugbau.* It is also considered to be the third of the three experimental prototypes ordered by *Ernst Udet* and was completed in 1940. The *Bv 141A V2* featured for the first time an asymmetrical tailplane to improved rear gunner's field of fire. The starboard half of the tailplane had been removed, while the post side tailplane was increased in length. *Dipl.-Ing. Walter Ballerstedt,* a staff test pilot with the *Erprobrungsstelle*-Rechlin who flew the asymmetrical machine, told this author in the mid 1980s that the *Bv 141A V2* showed no noticeable disadvantages in flight with its asymmetrical tail plane and was a rather pleasing machine to fly. It did, however, take the pilot only a short time to get accustomed to the strange-looking aircraft. Afterward, the pilots appreciated its easy handling and usefulness as a short-range reconnaissance and ground support machine. In fact, it had met all the *RLM's* specifications for a short-range reconnaissance and *Wehrmacht* ground support machine. Final disposition unknown.

Bv 141A V3 - *Werk Numer 359*, registered as *D-OLGA,* and radio call code *BL+AA*. This machine is considered by *Blohm & Voss Flugzeugbau* to be their *Bv 141B* production prototype and was completed in 1940. The *RLM* ordered five more prototypes, calling them pre-production prototypes based on the *Bv 141/A V3*, and are known as *Bv 141A-01* through *Bv 141A-05* and *werk numers 360* through *364.* All five of these so-called pre-production machines were handed over to *Erprobungsstelle*-Rechlin for testing one at a time as they came off the line. But at the same time the *RLM* continued to give the *Bv 141* a *Versuchs numer* (research number) *V4* through *V8.* The *Bv 141A V3* machine is reported to have handled well, *Walter Ballerstedt* told this author, although badly underpowered with its 865 take-off horsepower *BMW 132N* 9-cylinder, radial, air-cooled engine. *Ballerstedt* said, however, to improve directional stability even more, the *Bv 141A V3's* fuselage was lengthened in the five pre-production models from 37 feet 4 3/4 inches to 39 feet 10 1/3 inches, and its overall wing span was increased from 49 feet 6 1/2 inches to 50 feet 4 1/3 inches. To improve ground stability its wheel track was widened from 10 feet 1inches to 16 feet 10 3/4 inches. Offensive armament was installed for the first time in the *Bv 141A V3* and consisted of 2x*MG 17* 7.9 mm fixed forward-firing cannon and 2x*MG 15* 7.9 mm hand-held aft-firing cannon mounted in the rear of the *Bv 141A V3's* glazed nacelle cone. The hand-held aft-firing cannon could be rotated 360 degrees. Other equipment installed in the *Bv 141A V3* included an automatic camera, and racks for mounting 4x*SC50* kilogram [4x110 pound] bombs. Provisions for installing an *S-125* gas or smoke-spraying equipment were included in the *Bv 141A V3.* Inside the crew nacelle sat the pilot to port with the observer to starboard. The observer operated the bomb sight, camera, and radio equipment. The gunner lay prone over the wing's trailing edge. The *Bv 141A V3* also participated in bombing trials at Rechlin and Tarnewitz. Final disposition unknown.

Bv 141A V4 - *Werk Numer 360* (also known as *Bv 141A-01*), registered as *D-OLLE,* with radio call code *BL+AU.* Completed in early 1939 and intended for flight testing and overall evaluation by *Erprobungsstelle*-Rechlin. Wing span was increased to 50 feet 8 1/4 inches from 50 feet 4 1/3 inches. Wing area was in-

A nose-on view of a *Blohm & Voss Bv 141B*. Scale model and photographed by *Reinhard Roeser.*

creased, too, to 461.23 square feet from 457.46 square feet. Empty weight reduced to 6,845 pounds from 7,064 pounds, however, the loaded weight remained constant at 8,598 pounds. Initial flight testing by *Blohm & Voss Flugzeugbau* test pilots at Hamburg showed landing gear hydraulic problems, too, with this machine. It was severely damaged when entering its landing approach; one of the main wheels remained locked up in its wheel well, while the other was locked in the down position. Final disposition unknown.

Bv 141A V5 - *Werk Numer 361* (also known as *Bv 141A-02*) with radio call code *BL+AB*. Completed in January 1940 and turned over to *Erprobungsstelle*-Rechlin for flight testing involving low-level bombing, as it might be used for ground support operations with the *Wehrmacht*. At the end of its flight trials and evaluation Rechlin's test pilots gave it a favorable report in meeting the *RLM's* specifications. Nevertheless, the *RLM's OKL* (*Luftwaffe* high command) on April 4th, 1940, canceled its plans anyway for serial production, stating that it was underpowered with its 1x*BMW 132N* 9-cylinder, 865 takeoff horsepower, radial, air-cooled engine, and that no suitable higher horsepower power plant was available or would be in the foreseeable future. Final disposition unknown.

Bv 141A V6 - *Werk Numer 362* (also known as *Bv 141 A-03*) with radio call code *BL+AC*. Identical to the *Bv 141A V5*, except for a change involving a new upper circular defensive hatch which replaced the sliding hatch. Final disposition unknown.

Bv 141A V7 - *Werk Numer 363* (also known as *Bv 141 A-04*) with radio call code *BL+AD*. Identical to the *Bv 141A V6*. Final disposition unknown.

Specifications:

- Type - Three seat tactical short-range reconnaissance and army ground support aircraft.
- Engine - 1x*BMW 132N* 9-cylinder, radial, air-cooled engine rated at 865 takeoff horsepower for up to one minute's operation at sea level.
- Wing span - 57 feet 3 inches.
- Wing area - 462.5 square feet.
- Wing loading - NA
- Length, overall - 45 feet 10 inches.
- Weight, takeoff - NA
- Weight, empty - 9,025 pounds.
- Speed, maximum - 241 mph.
- Speed, cruising - NA
- Speed, landing - NA
- Crew - 3
- Range - NA

Bv 141A V8 - *Werk Numer 364* (also known as *Bv 141 A-05*) with radio call code *BL+AE*. Identical to the *Bv 141A V6*. Final disposition unknown.

A nose starboard view of a *Blohm & Voss Bv 141B*. Scale model and photographed by *Reinhard Roeser*.

Bv 141B V9 - (also known as *Bv 141B-01*), with radio call code *NC+OZ* and powered for the first time by the 1,560 take-off horsepower 1x*BMW 801A* 14-cylinder, radial, air-cooled "double-star" engine. It was found that the powerful *BMW 801A* was more than the aircraft structure could comfortably tolerate. Static vibration testing revealed that substantial increased strengthening of various components was required, including the tail assembly, undercarriage, and its control system. A tell-tale signature of the *Bv 141 V9* is its asymmetrical horizontal tailplane (elevator), which *Dr. Vogt* installed on a base over the rear fuselage to widen the field of cannon fire. The outer wing panels were also fitted with equi-taper, and its oval-shaped fuselage was changed to a circular shape. The main landing gear struts were changed. Prior to the change, the oleo strut formed an "L" shape where the wheel attached. With the change, the oleo strut was positioned directly over the wheel, and it was attached to the inverted "J," or fork. As a result of all the aerodynamic changes made, the *V9's* flight testing, which began on January 9th, 1941, its straight line speed was limited to a maximum of 280 mph. In addition, flight testing revealed aileron trimming to be ineffective, the ailerons themselves proved to be over-sensitive. With extensive modifications, *Bv 141B V9* was given over to *Erprobungsstelle*-Rechlin in May 1941 for further flight testing. Final disposition unknown.

Bv 141B V10 - (also known as *Bv 141B-02*) with radio call code *NC+RA* was delivered for service trials under service conditions to the *Aufklärungsschule #1* located at Grossenhain in Saxony. This was the second *Bv 141B* machine to be delivered to the *Aufklärungsschule #1*. Then, with the delivery of *Bv 141B-02*, *Blohm & Voss* was informed by the *RLM's General-Luftzeugmeister* (Quartermaster General) to provide sufficient *Bv 141Bs* to equip *Sonderstaffel Bv 141* for at least one operational squadron to be assigned to the Eastern Front (USSR), but these plans were canceled in the Spring of 1942 by the *RLM's* General Staff. It became clear that the *Bv 141B* would never receive the *BMW 801A,* plus the role for which the *Bv 141B* had been designed was now being handled on the Eastern Front by the twin *Argus As 410 V-12* engine of 430 horse powered *Fw 189A* from *Focke-Wulf*...and very effectively. It is believed that 860 *Fw 189As*, nicknamed *"Eule"* (owl), which when placed in service was replaced by the more correct name "Uhu" (eagle-owl), were produced, and it came to be called the *Das Fliegende Auge*, or "The Flying Eye" of the *Wehrmacht*. On the other hand, due to its inability to receive the *BMW 801A* engine, *Bv 141B* was destined never to see operational service. *Bv 141B V10* was transferred to *Aufklärungsschule # 1* during the Autumn of 1941. Final disposition unknown.

Headquarters and administrative center of *Blohm & Voss Flugzeugwerke,* Hamburg, Germany.

Specifications:

- Type - Three-seat tactical short-range reconnaissance and army ground support aircraft.
- Engine - 1x*BMW 801A* 14-cylinder, radial, air-cooled engine rated at 1,560 take-off horsepower for up to one minute's operation.
- Armament (defensive) - 2x*MG 17* fixed forward 7.9 mm cannon and 2x*MG 15* aft-firing 7.9 mm cannon on swivelable mountings.
- Armament (offensive) - 4x*SC 50* kilogram [110 pound] bombs on underwing racks.
- Speed, maximum - 229 mph at sea level and 272 mph at 16,400 feet.
- Range, normal - 745 miles.
- Range, maximum - 1,180 miles.
- Service ceiling - 32,810 feet.
- Weight, empty equipped - 10,362 pounds.
- Weight, normal loaded - 12,566 pounds.
- Weight, maximum - 13,448 pounds.
- Wing span - 57 feet 3 1/3 inches.
- Wing area - 569.41 square feet.
- Length, overall - 45 feet 9 1/4 inches.
- Height, overall - 11 feet 9 3/4 inches.

Bv 141B V11 - (also known as *Bv 141B-03*) with radio call code *NC+RC*. Delivered to *Erprobungsstelle*-Rechlin for flight evaluation. Final disposition unknown.

Dr.-Ing. Richard Vogt **(1894-1979), chief of design at** ***Blohm & Voss Flugzeugwerke*****, Hamburg. He earned his Ph.D. in aviation engineering from the University of Stuttgart in 1923 and was immediately hired by** ***Claudius Dornier*****.**

Bv 141B V12 - (also known as *Bv 141B-04*) with radio call code *NC+RF*. This machine was taken to Tarnewitz for armament testing. Its performance was disappointing. It was discovered that its cannon troughs were too short and that cordite fumes filled the cockpit when the cannon were fired. Final disposition unknown.

Bv 141B V13 - (also known as *Bv 141B-05*) with radio call code *NC+RH*. It is thought that this was the very last *Bv 141* built, and it was delivered to *Erprobungsstelle*-Rechlin for evaluation on May 15th, 1943. Final disposition unknown.

Bv 141B - *Werk Numer* unknown, appears to be a *B*-series type and *"V"* number unknown, but with partial radio call code *(G)K+GH*. Found by American Army severely damaged (severed tail assembly and all cockpit cabin glass knocked out) probably due to a nearby bomb blast. There a further mystery. Although it appears to be a *B*-series type, it has the horizontal tail plane of the *A*-series type. Since this machine was found on a remote *Luftwaffe* air base among other *Luftwaffe* machines, perhaps the tail assembly from a wrecked *A*-series type was dropped off near the *Bv 141B's* severed tail.

No *Blohm & Voss Bv 141s* are known to have survived the war.

Other Proposed Asymmetrical Aircraft Designs From *Blohm & Voss*

Dr.-Ing. Richard Vogt and his design staff at *Blohm & Voss Flugzeugbau* completed at least eleven known aircraft designs with an asymmetrical wing planform in addition to their *Bv 141* between early 1938 and May 1945. Perhaps even more. Several have mixed power plant arrangements, that is, containing a piston engine and a turbojet engine. One of the designs called for only a single turbojet engine. It is believed that none of the eleven designs were accepted by the *RLM*, and no contacts written for the construction of prototypes. Additional information about these several asymmetrical designs can be found in my book *Secret Aircraft Designs of the Third Reich*, Schiffer Publishing Ltd., Atglen, Pennsylvania, 1998, ISBN: 0-7643-

Ernest Udet**. He was in charge of aircraft planning and development within the** ***RLM*****. He was so impressed with the flying qualities of** ***Vogt's Blohm & Voss Bv 141A*** **after a test flight in February 1938, that he ordered 500 of the machines be constructed.**

Erhard Milch.

Reichsmarschall Hermann Göring. Shown in Summer 1940 along the French coast facing England. *Göring* was there to visit and motivate *Luftwaffe* formations stationed along the French coast.

Richard Vogt, taken prior to his death (January 23rd, 1979) at Santa Barbara, California.

0564-6. E-mail: Schifferbk@aol.com. The author can be reached at: Myhrarn@Juno.com.

Bv P.111 - Use: A 3-engined asymmetrical sea plane.

Bv P.163.01 - Use: General Purpose Aircraft
This single-engine aircraft had a most unusual layout, because the cockpit cabin and crew cabin were mounted on each wing tip. It was to be powered by a 3,800 take-off horsepower *Daimler-Benz DB 613C* piston engine mounted in the nose of center or normal fuselage. Fuel and bombs would have also been carried in this center fuselage. The port nacelle housed the pilot and radio operator/observer, and is fitted with 2x*MG 151* cannon in a movable twin mounting, firing astern, and containing 500 rounds per cannon. The starboard nacelle carries two cannon operators who are provided with 1x*MG 151* cannon on a movable mount, firing forward and carrying 250 rounds. In addition, the proposed machine was to have carried 3x*MG 151* in a dorsal turret. One *MG 151* firing forward with 500 rounds and 2x*MG 151* firing astern with both cannon containing 500 rounds each. Its *DB 613C* 24- cylinder piston engine would have driven a contra-rotating propeller. The *Bv P.163.01,* an idea for a proposed bomber although not an asymmetrical planform, the *P. 163.01* did have wing tip nacelles for pilot and crew with a center fuselage carrying fuel and bombs, and therefore is considered asymmetrical in planform. A wing tip nacelle containing pilot and crew was tested on a *Bv 141* to gain actual experience of how this idea might work for the pilot and crew. Unfortunately, this author has no photo showing the modified *Bv 141* with its wing tip crew nacelle.

Specifications:

- Wing span - 68 feet.
- Wing area - 595 square feet.
- Wing loading - 55.5 pounds per square foot.
- Aspect ratio - 6.57.
- Length, overall - 51 feet.
- Engine - 1x*Daimler-Benz DB 613C* rated at 3,800 take-off horsepower for up one minute's operation.
- Weight, maximum flying - 33,000 pounds.
- Weight, fuel - 6,000 pounds.
- Speed, at climb with maximum weight at sea level - 276 miles per hour.
- Speed, at climb with maximum weight at 19,759 feet - 338 miles per hour.
- Rate of climb, maximum weight at sea level - 1,240 feet per minute.
- Rate of climb, maximum weight at 19,570 feet - 810 feet per minute.
- Ceiling, maximum weight of 33,000 pounds - 27,800 feet.
- Ceiling, absolute - 29,600 feet.
- Take-off distance on grass - 1,600 feet.
- Take-off distance on concrete - 1,480 feet.
- Landing speed - 87 miles per hour.
- Range, cruising at sea level - 1,280 miles.
- Range, cruising at 19,750 feet - 1,500 miles.
- Bomb load - 4,400 pounds.

***Claudius Dornier* on a flight with a *Dornier* company pilot. He was a mentor to *Richard Vogt*, as well as numerous other German aircraft designers from the 1930s.**

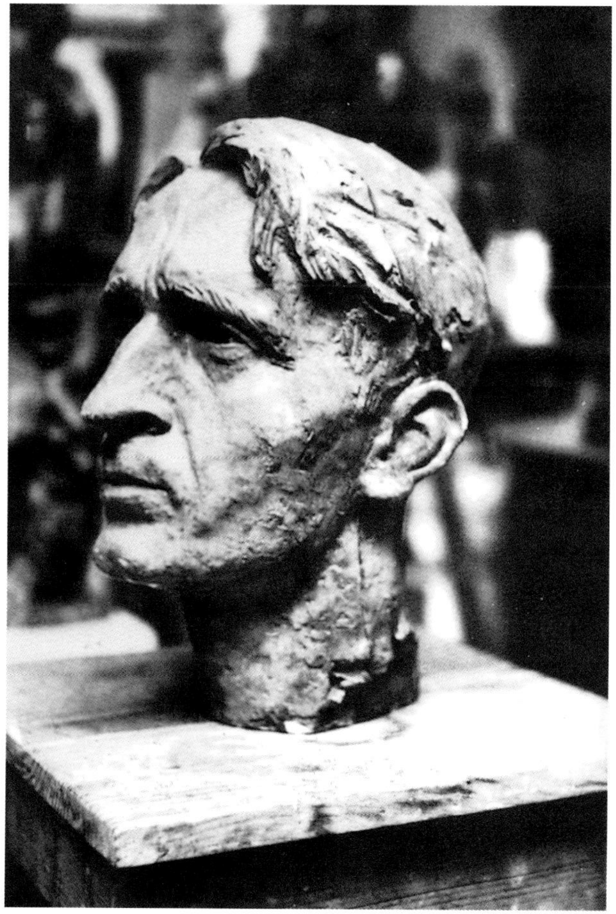

A bronze metal casting of *Richard Vogt*.

Bv P.163.02 - Use - General Purpose Aircraft
Same general layout as its sister *Bv P.163.01*, however, it would have been powered by 1x *BMW 803A* 28-cylinder piston engine driving a contra-rotating propeller.

Bv P.170.01 - Use - NA

Bv P.177 - Use - Multi Purpose Dive bomber and Ground Attack Aircraft

This asymmetric dive bomber and ground attack aircraft was to have been powered by a *Junkers Jumo 213* piston engine mounted in the nose of the fuselage. Its wing, apart from bulges to accommodate the main wheels which project forward at the tips, was of a constant chord design. It would have had a wing span of 39.3 feet and a wing area of 258 square feet. It was offset in relation to the fuselage. Its tailplane, however, was symmetrical. Accommodation for the crew of two was to have been provided in a nacelle on the starboard wing. The pilot and the navigator/observer/rear gunner sat back to back. This dive bomber was to be also equipped with two forward-firing and two rearward firing cannon, all located in the nacelle on the starboard wing tip. Additional armament mentioned were 2x *MK 103* cannon slung under the wing's leading edge and a bomb carrier directly beneath the nacelle.

The *Bv P.177* asymmetrical design was put forward by *Dr. Vogt* in 1942 as a proposed single seat dive bomber. It would have been constructed mainly of wood, and was to have a similar planform like the *Bv 141.* It would have been powered by a 1,560 take-off horsepower for up to one minute's operation *BMW 801A* 14-cylinder radial air-cooled engine. An order was extended by the *RLM*, and the prototype was designated *Bv P. 237.* The project was canceled because the demand for the *BMW 801As* from the *Focke-Wulf Fw 190* fighter and the *Dornier Do 217* twin-engine bomber was greater than *BMW* could supply.

Bv P.178 - Use - NA

Bv P.179 - Use - NA

Bv P.194.01 - Use - *Wehrmacht* ground support and dive bomber

This was one of *Blohm & Voss'* four proposals in a single package in response to the *RLM's* February 1944 call for a new single-seat ground-support and dive bomber to replace the aging and slow-moving *Junkers Ju 87 Stuka* dive bomber. They included the *Blohm & Voss P.192.01*, *P.193.01*, *P.194.01*, and *P.196.01. Bv's P.194.01* (their third entry out of a package of four submitted) was the only machine in this proposal to have an asymmetrical wing arrangement...similar to their earlier piston-powered *Bv 141* tactical short-range reconnaissance machine.

A poor quality pen and ink drawing from *Blohm & Voss* of their *Bv 141's* three man crew cabin as seen from its port side and featuring its internal structure.

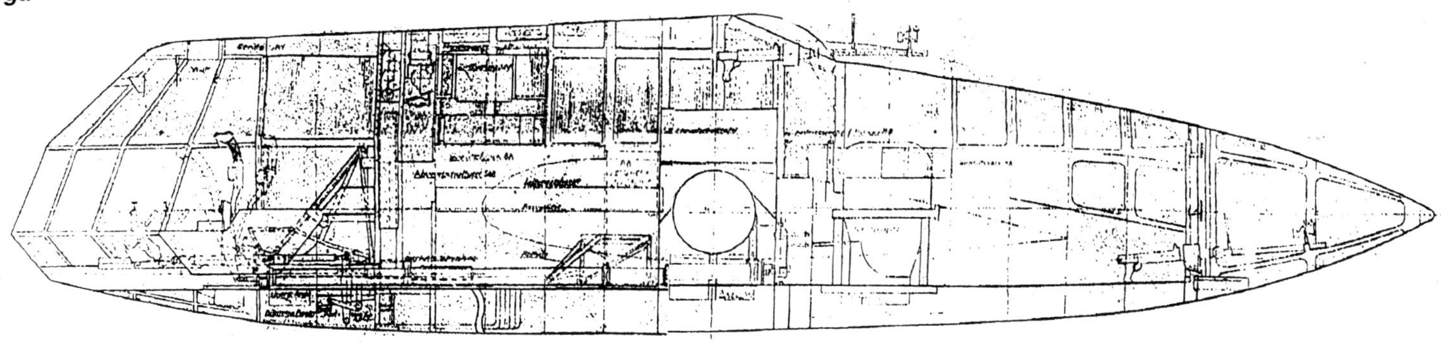

The mainly all wood *Bv P.194.01* was proposed to carry one *Junkers Jumo 004* turbojet engine mounted on the under surface of the pilot's gondola in addition to a fuselage nose-mounted *BMW 801D* radial piston engine having 2,200 takeoff horsepower. Had this machine been built as *Blohm & Voss* had hoped, their *Bv P.194.01* would have been the world's first asymmetrical turbojet-powered aircraft. Another engine possibility *Dr. Vogt* was anticipating was the *BMW 018* turbojet engine. The *BMW 018* was expected to be Germany's most powerful turbojet engine with a thrust rating of 7,525 pounds [3,402 kilograms]. *Vogt* believed that given the 7,525 pounds thrust from the *BMW 018* on his *Bv 194.01* that it would completely cancel out the torque coming from the powerful *BMW 801D* 2,200 horsepower piston engine. None of the four proposals in this *Blohm & Voss* package were successful.

Specifications:

- Wing span - 50 feet 2 inches.
- Length, overall - 38 feet 9 inches.
- Engine - 1x*BMW 801D* and 1x*BMW 003* or the *BMW 018* turbojet when it became available for field use.
- Speed, cruising - 373 mph [600 kph] at sea level.
- Speed, maximum - 450 mph [725 kph] 23,620 feet [7,200 meters].
- Service ceiling - 35,500 feet attitude.
- Range - 620 miles [1,000 kilometers]
- Armament - 2x*MG 151* cannon and 2x*MK 103* all mounted in the fuselage nose.
- Bomb load - 9x*SC 70* or 2x*SD 500.*
- Weight, maximum takeoff - 24,690 pounds [11,200 kilograms].
- Project abandoned in 1944.

Bv P.194.03 - Use - NA

Bv P.194.10 - Use - NA

Bv P.237 - Use - NA
This was to be a *Bv 141B,* but with the addition of a single turbojet engine, either a *BMW 003A* or a *Junkers Jumo 004B.* Three-view drawings of this proposed machine feature the 1,984 pound [900 kilogram] thrust *Jumo 004* turbojet engine, for example, mounted on the under surface of the starboard wing...between the *2,200* takeoff horsepower *BMW 801D* 14-cylinder radial, air-cooled engine and the crew nacelle. Estimated top speed was 452 mph [727 km/h] at 22,967 feet [7,000 meters] altitude decreasing to 397 mph [638 km/h] at sea level. An operational radius of 630 miles [1,041 kilometers] was anticipated, with the aircraft carrying a normal complement of bombs and ammunition. The proposed *Bv P.237* with its mixed power plants of one piston and one turbojet engine was canceled by the *RLM* due to the unavailability of the *BMW 801...*just as the *Bv 141* machine had been years earlier.

Specifications:

- Wing span - 47 feet 5 1/4 inches [14.46 meters].
- Length, overall - 35 feet 3 1/4 inches [10.75 meters].
- Weight, takeoff -14,740 pounds [6,685 kilograms].
- Speed, maximum - 357 mph [575 km/h].
- Range -1,642 miles [2,000 kilometers].

Bv P.237.01 - Use - NA

Bv P.237.02 - Use - NA

Bv Ae 607 - Use - possible day fighter. Fact or fiction? The authenticity of this proposed *Heinkel-Hirth HeS 011* turbojet powered delta wing project has not yet been proven, however, the pilot's cockpit is sometimes shown to port...other times to starboard. It frequently appears as a tailless machine with only a rudder for a tail. In addition, it is supposed to have had a sweep to its wings and two features which make it unique among German aircraft projects: nose-mounted canards (port and starboard), which are thought to have served as elevators; and second, its

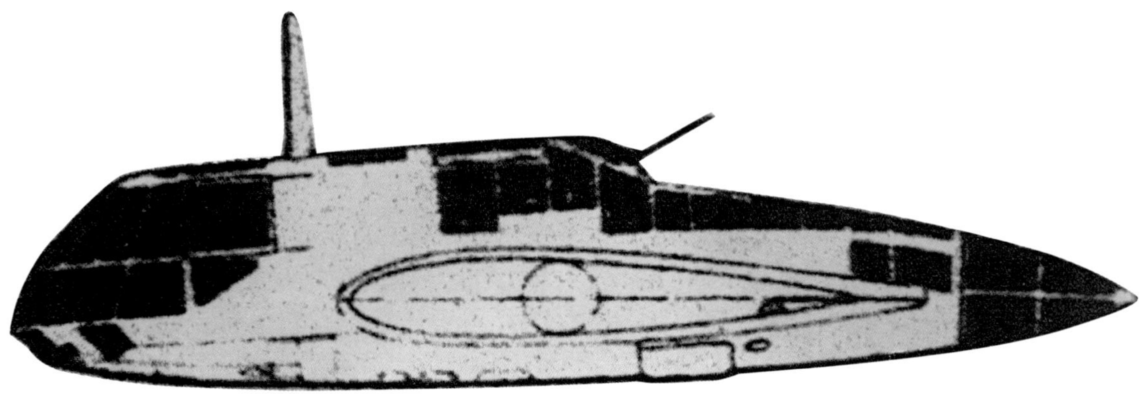

A poor quality pen and ink drawing from *Blohm & Voss* of their *Bv 141's* cockpit cabin featuring its glass-window areas (black) and metal surfaces (white). The wing profile is shown in the center of the drawing. The round circle is the wing's tubular main spar.

single seat cockpit was off center, usually far to the port side of the machine. The aircraft's landing gear was to have consisted of four main landing wheels. The two front wheels were to be steerable and retract to the outside under the surface of the fuselage wings. The rear wheels, which were to bear the larger percentage of weight, were to retract into the rear trailing edge of the fuselage wings. No estimated performance data is available on this type...indeed, if it were a *Blohm & Voss* type at all.

Specifications:

- Engine - 2x *Heinkel-Hirth HeS 011* turbojet units each having 2,866 pounds [1,300 kilograms] thrust.
- Crew - 1.
- Armament - 3x *MK 108* 30mm cannon.

* The *Blohm & Voss Flugzeugbau* (aircraft company) was originally founded July 4th, 1933, as the *Hamburger Flugzeugbau GmbH* (a division of the *Blohm & Voss Schiffswerft* (shipbuilding company). Its aircraft carried the designation *Ha*, given by the *RLM* for *Hamburger Flugzeugbau.*

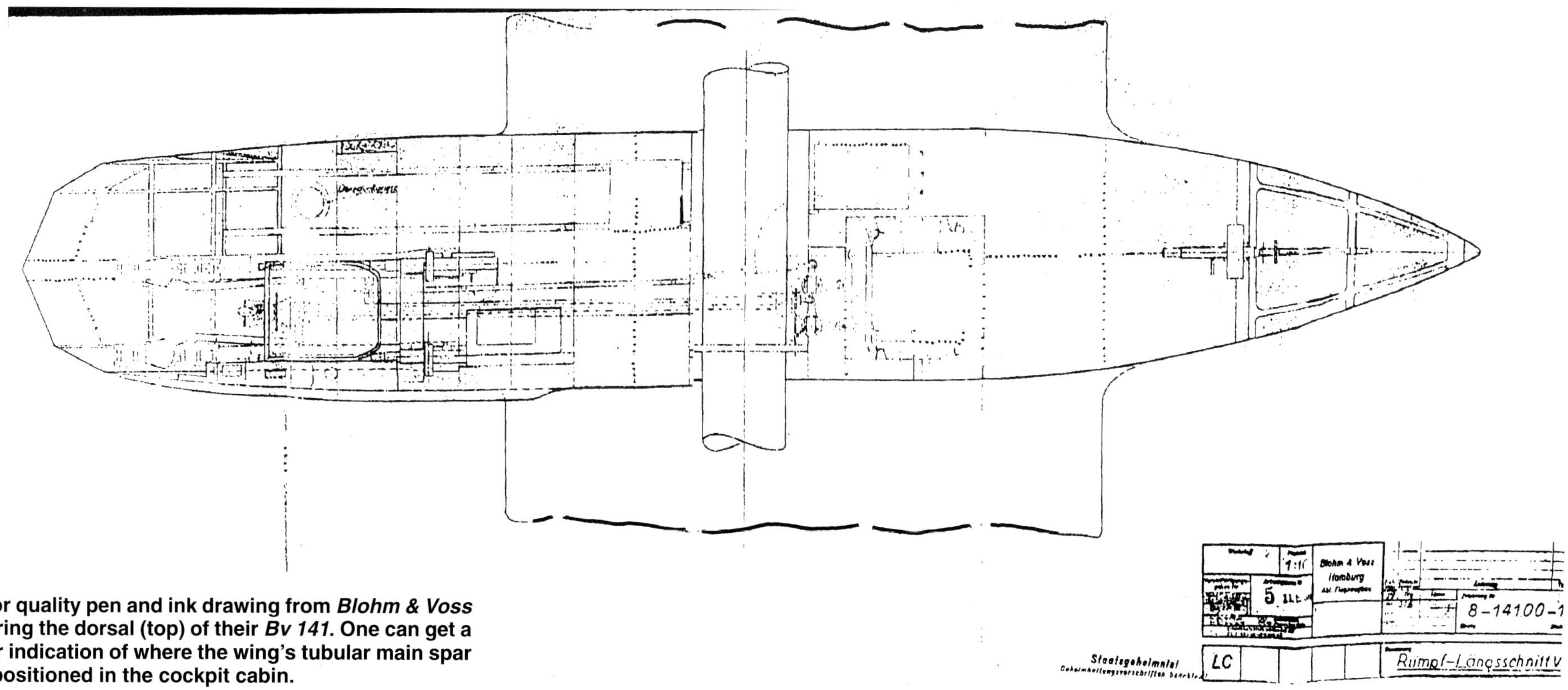

A poor quality pen and ink drawing from *Blohm & Voss* featuring the dorsal (top) of their *Bv 141*. One can get a better indication of where the wing's tubular main spar was positioned in the cockpit cabin.

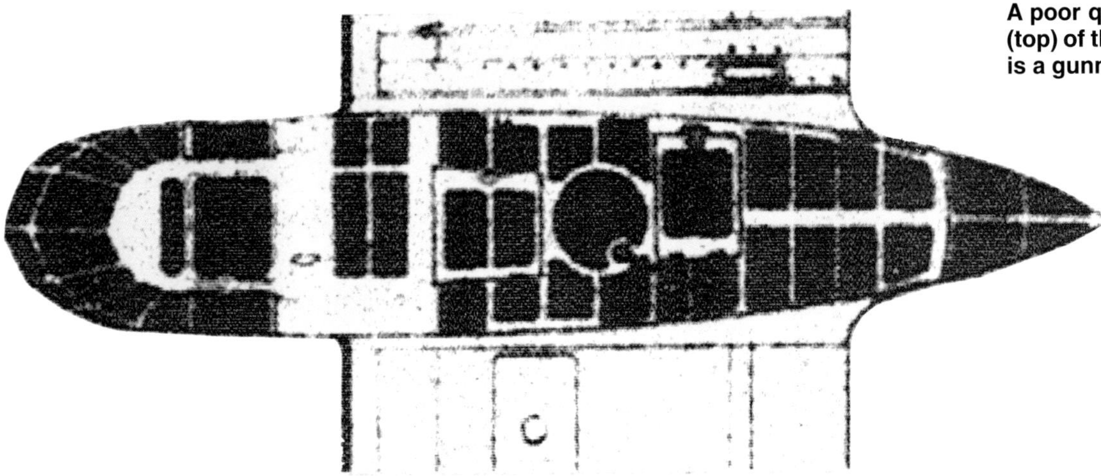

A poor quality pen and ink drawing from *Blohm & Voss* featuring the dorsal (top) of their *Bv 141's* glassed in areas. The round circle seen about mid-wing is a gunner's turret.

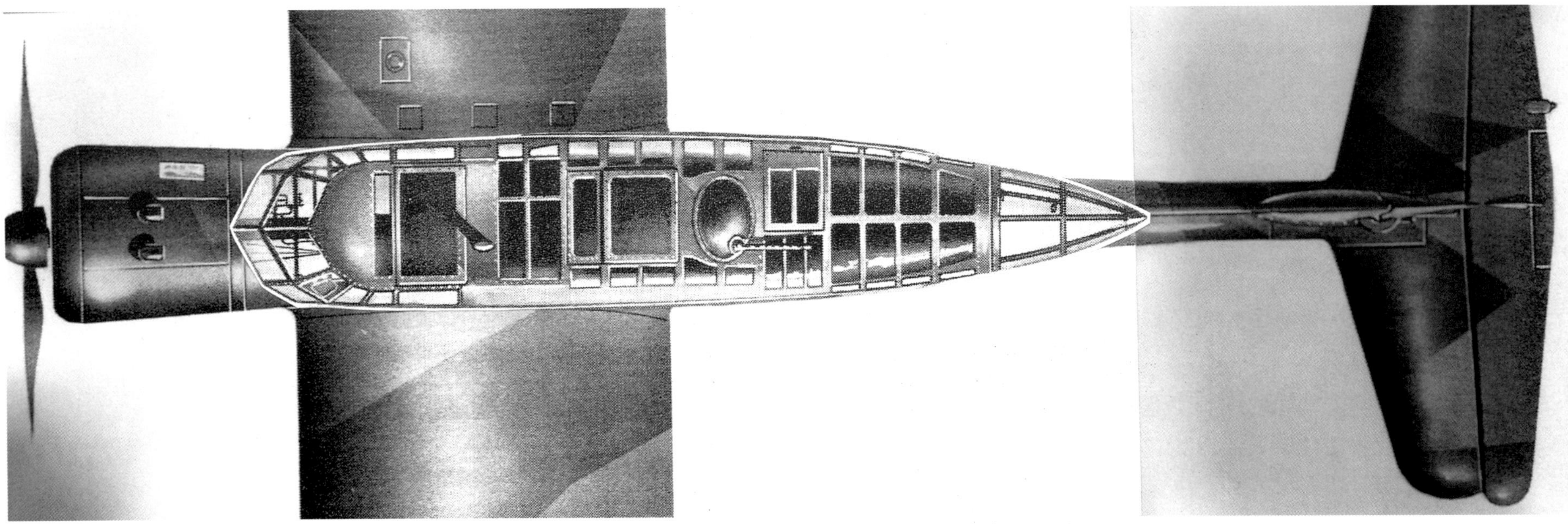

RLM's specifications for a new *Wehrmacht* short-range reconnaissance, ground support, and light bomber called for a three man crew...pilot, bombardier, and rear gunner. Furthermore, *RLM's* specifications stated that the three man crew should have excellent visibility. To accomplish these goals, *Dr. Vogt's* initial thinking was to place the three man crew's cabin on a standard fuselage, behind the engine. *RLM's* specifications also included that the aircraft be powered by a single engine. This is pretty much how a three man crew's cockpit cabin would have looked...poor reconnaissance viewing, as well as for providing ground support and light dive bombing. *Vogt* realized that this arrangement would not work and another layout would be required. Composite photo made from images of the *Blohm & Voss Bv 141B* by *Gert W. Neumann*.

Arado flugzeugbau's entry (*Ar 198*), along with the *Blohm & Voss Bv 141A* and the *Focke-Wulf Fw 189* into *RLM's* competition. Notice that *Arado* included the three man crew in the fuselage of their high braced wing. One can only imagine what *Arado's* engineers were thinking when they submitted this obsolete design for consideration as a short range reconnaissance and *Wehrmarcht* ground support machine. It was rejected immediately by the *RLM*, and the competition thereafter was only between *Blohm & Voss* and *Focke- Wulf*.

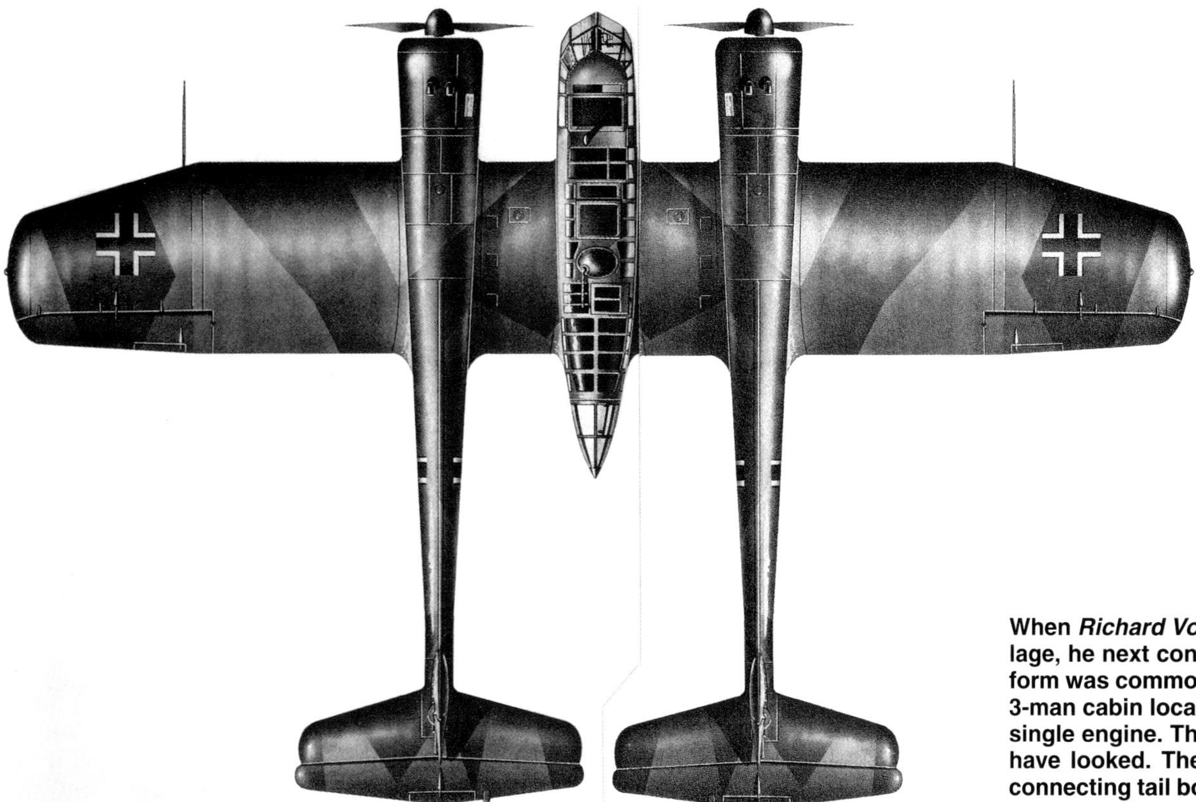

When *Richard Vogt* threw out the idea of placing the 3-man cabin on the fuselage, he next considered placing the cabin between two fuselages. This planform was common, indeed, *Blohm & Voss'* competition *Focke-Wulf* solved the 3-man cabin location this very way. Yet, *RLM's* specifications called only for a single engine. This is probably how a twin boom *Blohm & Voss Bv 141* might have looked. The elevators probably would have been a one piece, that is, connecting tail booms. Composite photo made from images of the *Bv 141B* by *Gert W. Neumann*.

Richard Vogt's answer to the *RLM's* need for a 3-man cabin and powered by a single radial engine...his asymmetrical planform *Blohm & Voss Bv 141*. Asymmetrical planforms would become *Vogt's* signature. Courtesy of *Gert W. Neumann*.

Asymmetrical flying machines were not unheard of when *Richard Vogt* suggested it to the *RLM* in late 1937. Shown here is a pen and ink drawing of the asymmetrical "*Gotha*" bomber of *WWI*. *Goether Waggonfabrik*-Gotha built the physically largest aircraft of *WWI*, and because they were used to bomb England, they were the only aircraft manufacturing company in Germany that was completely torn down to the ground after the war.

An asymmetrical *Heinkel He 116* prototype civil passenger airplane with two motors on its starboard wing and a single engine on the port wing? No. The *He 116 V2* "*Hamburg*" *D-AJIE* lost its outer port wing in the area of Leipzig-Brandis in 1938. The machine was on route to Rostock and made an emergency landing at Leipzig-Brandis. The engine could not be repaired, so it was decided to remove the entire remains of the engine and cover the hole over with aluminum and then proceed on to Rostock, which it did.

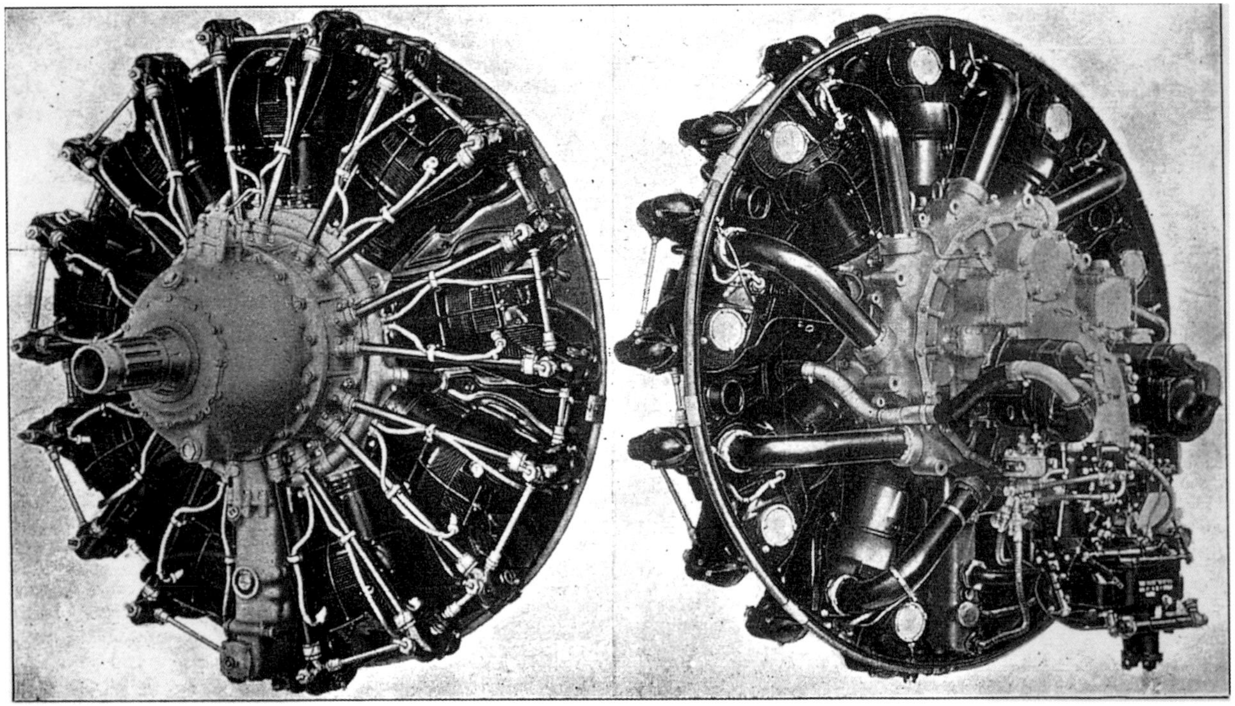

Each of the *Blohm & Voss Bv 141A* prototypes had to use the 865 takeoff horsepower *BMW 132N* 9-cylinder air cooled radial aeroengine. It was rated at only 665 horsepower at 14,760 feet [4,500 meters]. Very inadequate. Left: forward or engine front. Right: Aft or engine backside.

The starborad side of the *BMW 801A* 14-cylinder "double star" air cooled radial engine of 1,565 takeoff horsepower. *Dr. Richard Vogt* was able to install several of these high horsepower engines in his *Blohm & Voss Bv 141B*, however, he would never be able to have for series production...the reason is that the *Focke-Wulf Fw 190* and the *Dornier Do 217* required the same engine, and *BMW* could not make enough for all three...barely enough for the *Fw 190* and *Do 217*. Left: aft. Right: forward featuring the radiator fan.

A pen and ink drawing of a starboard side *BMW 801N* as it would appear installed, including its full cowling.

The winner of *RLM's* competition for a new short range reconnaissance and *Wehrmacht* ground support machine...the *Focke-Wulf Fw 189A* shown here from its rear starboard side and nearing completion.

A *Focke-Wulf Fw 189A* in flight as seen from its nose port side. Although *RLM's* specifications called for a single engine, *Kurt Tank* of *Focke-Wulf* was able to get by with two *Argus* in-line engines. Notice, too, the similarity of the *Fw 189A's* cockpit cabin and that of the *Blohm & Voss Bv 141A.*

A highly unusual photograph of a *Blohm & Voss Bv 141B* from the cockpit as it banks into a hard turn to port. Notice the *Bv 141B's* sistership out in front and the German countryside beneath.

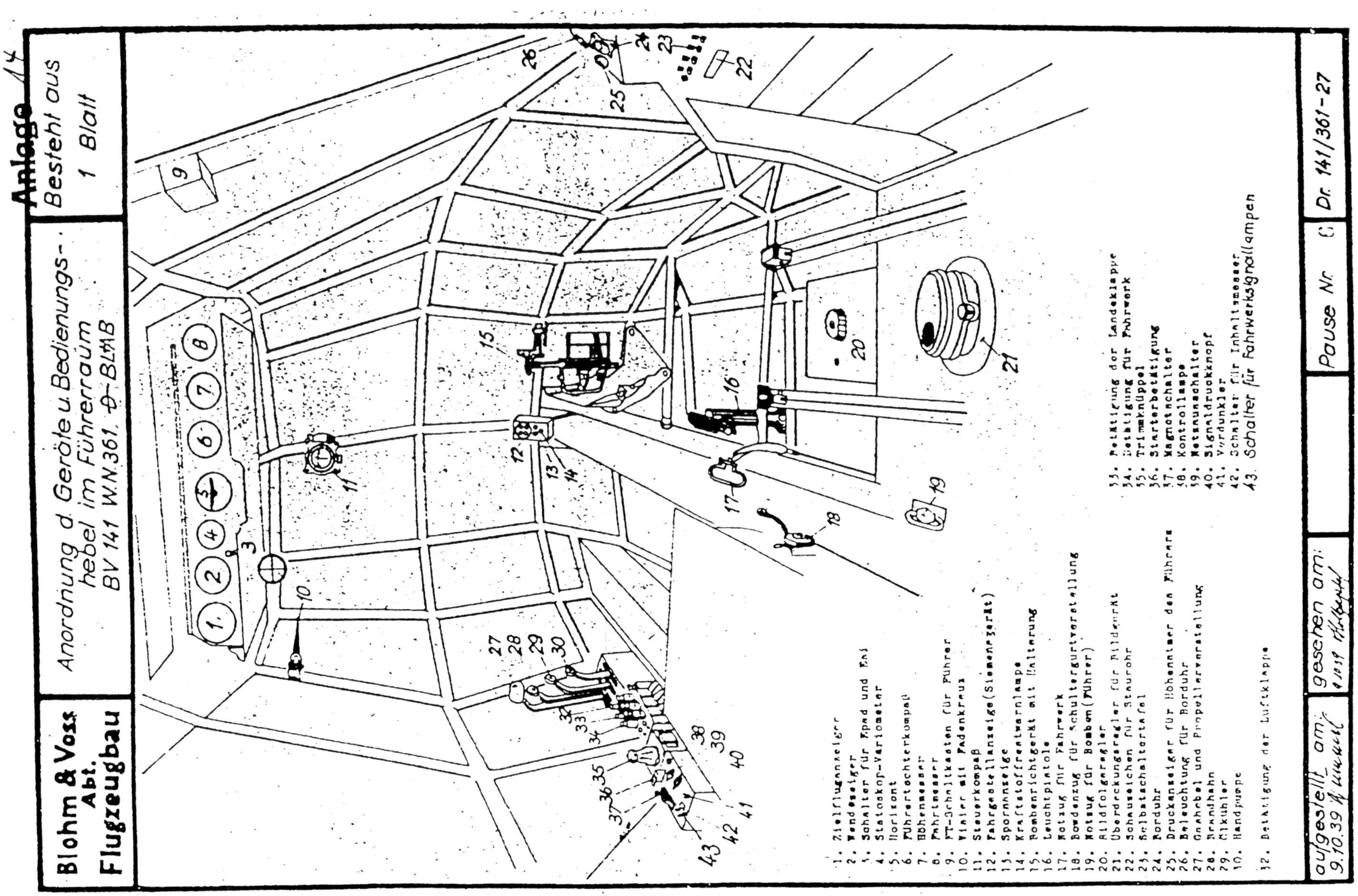

A pen and ink illustration by *Blohm & Voss* of their *Bv 141A, werk numer 361, D-BL+AB* cockpit as seen from behind and starboard. The illustration is dated October 9th, 1939. The pilot's seat is off to port.

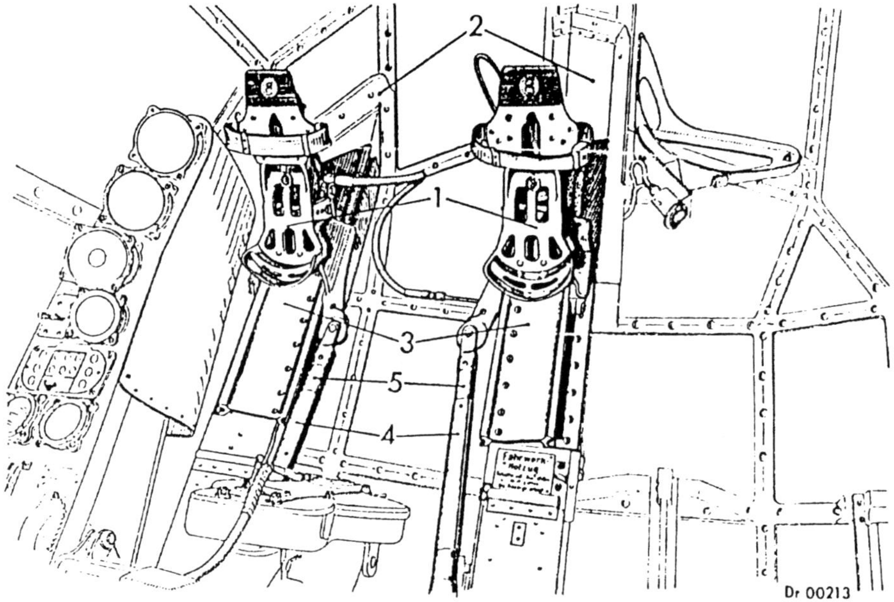

1 = Fußhebelpaar
2 = MG-Schußkanäle
3 = Preßstoffschienen
4 = Verbindungsmuffe
5 = Stoßstangen

Abb. 3: Fußsteuerhebel

A pen and ink illustration from *Blohm & Voss* featuring the pilot's rudder control foot pedals in a *Bv 141*. The pilot's flying position was from the port side of the cockpit.
01 - dual foot lever
02 - machine gun cannon troughs
03 - rudder control push rods
04 - rudder control push rod ends

Right: A pen and ink drawing from *Blohm & Voss* featuring the sector panel for the machine's landing flap control linkage in the *Bv 141B*:
01 - dual pressurized cylinders
02 - piston rod
03 - reverse lever
04 - activating lever
Qr - aileron control linkage
Hr - elevator control linkage
Sr - rudder control linkage
L - landing flap control linkage

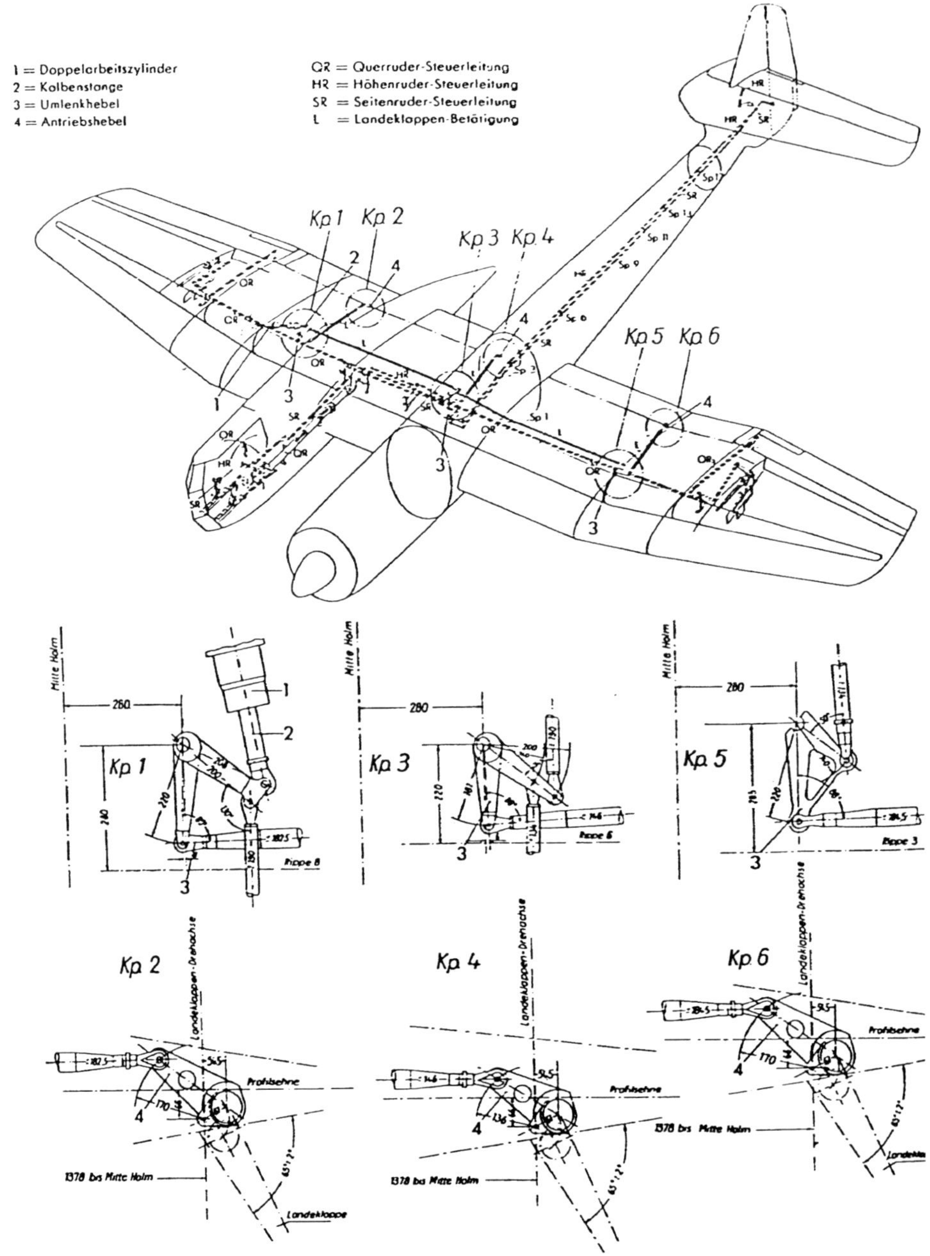

Kp 1—6 in „0"-Stellung gezeichnet

Zeichnung 6: Einstellplan der Landeklappensteuerung

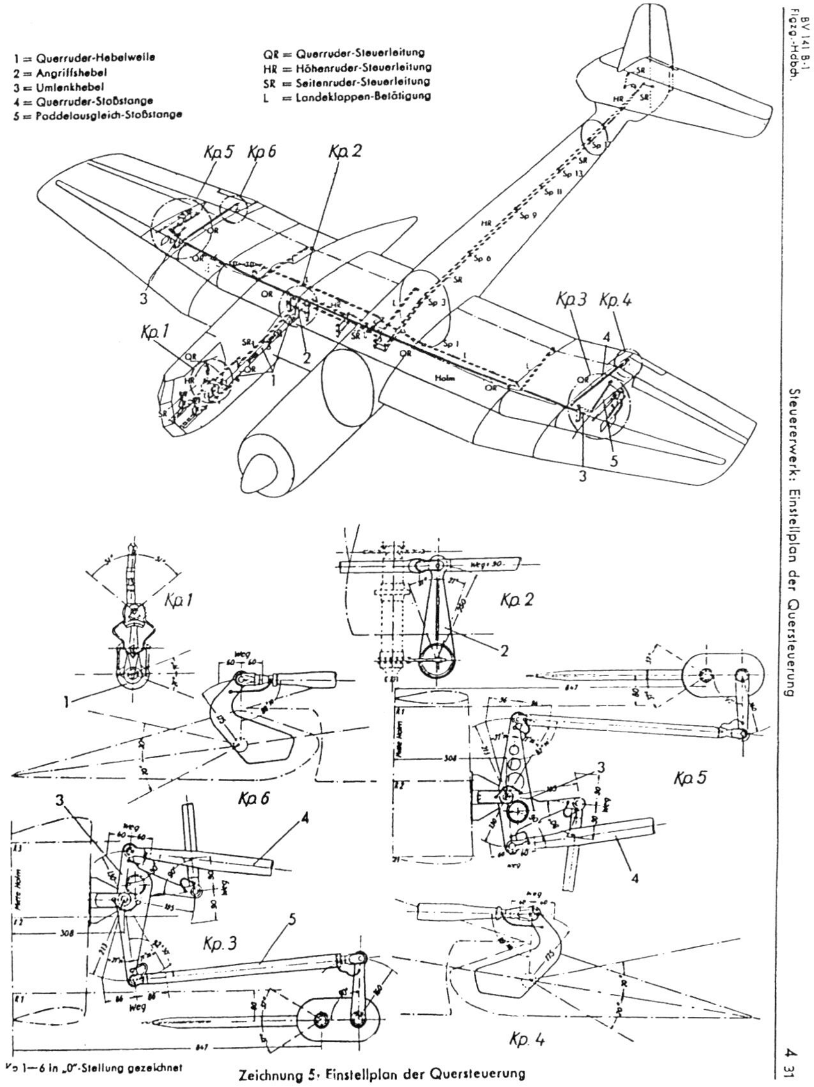

A pen and ink drawing from *Blohm & Voss* featuring aileron control linkage in their *Bv 141B*:

01 - aileron lever axle
02 - activating lever
03 - reverse lever
04 - aileron control/cable linkage
05 - foot paddle/compensation push rod

Qr - aileron control/cable linkage
Hr - elevator control/cable linkage
Sr - rudder control/cable linkage
L - landing flap control

The fuselage of a *Blohm & Voss Bv 141* under assembly featuring the lower portion of the fuselage. Notice the tail plane assembly control linkage running along the inside bottom.

The *Blohm & Voss Bv 141* wing's were constructed around a single tubular metal spar. The spar in turn was bolted to a "T" bone-type structure built in to the fuselage. To the left of the photo is the lower fuselage of a *Bv 141*, and the round pipe-like protrusion is where the port side wing's tubular spar would be bolted on.

An inboard section of a *Blohm & Voss Bv 141's* port wing containing the port main wheel assembly (located between the cockpit cabin and the outboard wing). Notice the port main landing gear and wheel well seen to the left in the photograph.

A starboard outer wing section from a *Bv 141*. Notice its tubular metal spar running the length of this outboard section and the very thin ribs.

Richard Vogt conceived the idea of using a large diameter (775 mm) metal alloy pipe as the wing spar. Although the tubular spars shown in this photo are from the *Hamburger Ha 138*, they are similar to those *Vogt* placed in his asymmetrical *Blohm & Voss Bv 141.* Notice the finished tubular spar at the right...it has small metal clips welded on at the exact position where the ribs would be attached.

BV 141 B-1
Flgzg.-Hdbch.

Elektrisches Bordnetz: Beschreibung

9 B 27

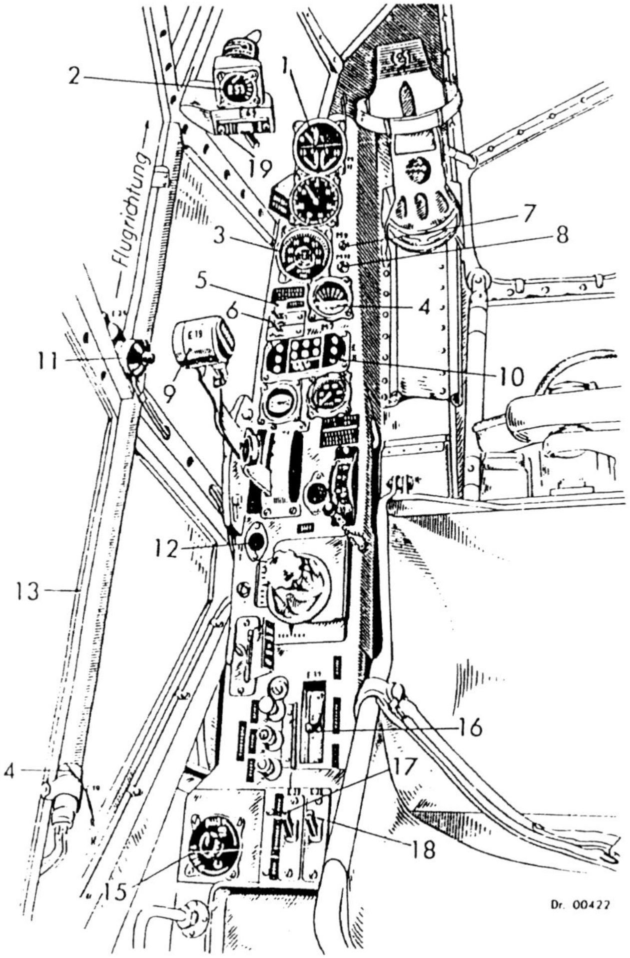

1 Vierfachanzeigegerät
2 Wegmesser (Querrudertrimmung)
3 Drehzahlmesser
4 Kraftstoffvorrats-Anzeigegerät
5 Schalter für Kraftstoff Förderpumpe des vorderen Behälters
6 Schalter für Kraftstoff-Förderpumpe des hinteren Behälters
7 Merkleuchte für Reststandwarnung
8 Merkleuchte für Reststandwarnung
9 Daumenschalter f. Luftschraubenverstellung
10 = Zwölflampengerät
11 = Unterbrecherknopf (Boschhorn)
12 = Signalknopf (H 1)
13 = Leuchtstab
14 Lampe für Leuchtstab
15 = Verdunkler
16 = Fahrwerksschalter
17 = Schalter für Drosselring
18 = Schalter für Nasenspaltring (Ölkühler)
19 = Umschalter für Luftschraubenverstellung

Abb. 18: Schalter und Schauzeichen der elt. Antriebe auf der Gerätebank

B 141 B 1
Flgzg.-Hdbch.

Schußwaffenanlage: Kurzbeschreibung

8 A 11

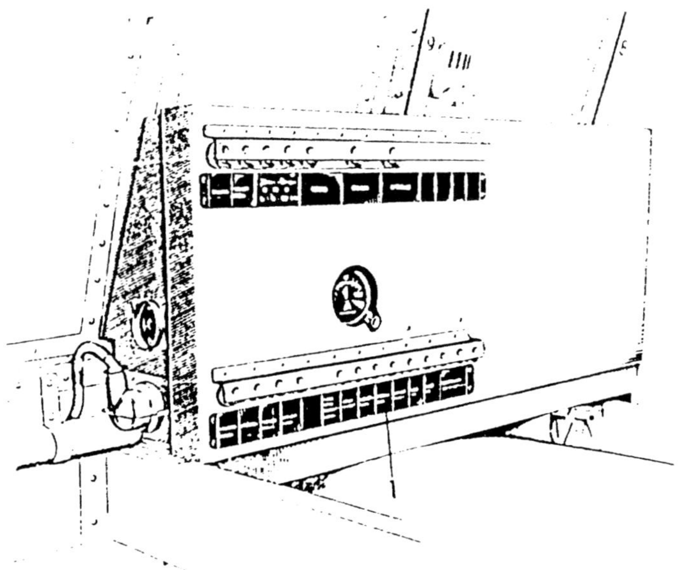

1 Selbstschalter „Starre Waffen"

Abb. 4: Hauptschalttafel

1 = Sicherungsschalter für MG 17

Opposite, Left: Switches and dials found on the *Bv 141's* instrument panel:

01 - artificial horizon
02 - aileron control position
03 - engine revolutions per minute
04 - fuel quantity gauge
05 - forward fuel tank pump switch
06 - aft fuel tank pump switch
07 - low fuel warning indicator (light)
08 -
09 - propeller pitch control thumbswitch
10 - 12 amp gauge
11 - circuit breaker
12 - signal knob
13 - electric torch
14 - light for electric torch
15 - light dimming switch
16 - landing gear switch
17 - electrical switch
18 - oil cooler switch
19 - propeller pitch control electric switch

Opposite, Right:

Upper - 01 - pilot switch control panel for the fixed cannon.
Lower - 01- safety switch for the *MG 17* cannon.

A *Blohm & Voss Bv 141B* making a very close encounter to the camera airplane. Notice the three other *Bv 141B* sister ships seen at the top of the photograph.

The two *Blohm & Voss Bv 141Bs*, now perhaps wing tip to wing tip, while the cameraman captures the moment.

Our *Blohm & Voss Bv 141B* completes its close encounter and is seen moving away to port.

A *Blohm & Voss Bv 141B* in a climb shortly after takeoff from *Erprobungsstelle*-Rechlin. This angle provides a good view of its asymmetrical horizontal tail plane.

A poor quality photo of a *Bv 141B* as seen from its nose port side. It appears that this machine is occupied only by its pilot.

Walter Ballerstedt, Möne, West Germany, 1985. Photo by author.

This is how the British aviation magazine "*Aeroplane*" viewed *Richard Vogt's* asymmetrical creation...not very kindly, it appears.

The *Blohm & Voss Bv 141A V1* as seen from its port side and carrying its civil registration *D-OTTO*. Photographed during its September 1938 initial flight testing.

An overhead view of ***Blohm & Voss' Bv 141A V1 D-OTTO*** **as seen from its port side.**

The *Blohm & Voss Bv 141A V1* prototype was to have been equipped with four cannon...one rear-firing swiveling single *MG 15* 7.9 mm cannon. To fire the *MG 15*, the gunner would have first had to open the rear hatch as shown.

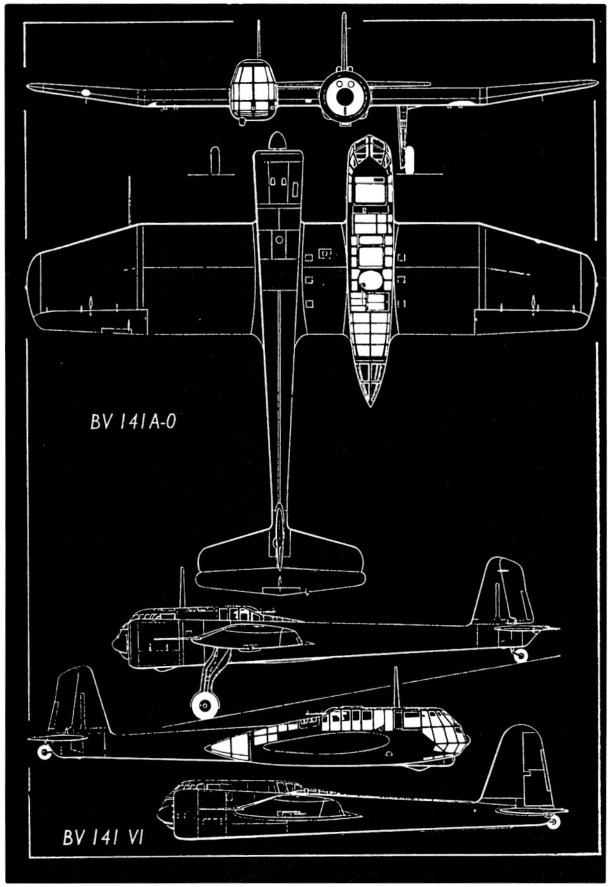

A pen and ink 3-view illustration of the *Blohm & Voss Bv 141A V1*, the so-called "official" prototype *D- OTTO.*

Hydraulic failure on the *Blohm & Voss Bv 141A V1 D-OTTO* required a forced landing in a ploughed farm field in October 1938. Its port landing gear extended (or half extended), but the starboard did not extend beyond its wheel well.

A port side view of *Blohm & Voss Bv 141A D-OTTO* (*V1*) October 1938. Its port wing was undamaged, however, its starboard wing suffered considerable damage.

Above: A ground level view of the *Blohm & Voss Bv 141A V2* (*Ha 141-0*) civil registration *D-ORJE*. Winter 1938.

The *Blohm & Voss Bv 141A V2,* as seen from its starboard side, on its maiden flight February 25th, 1938.

The *Blohm & Voss Bv 141A V2*. Notice that the forward glazed section of the cabin has a bubble allowing the observer a good view side to side and below.

A starboard side close up of the glazed cockpit cabin nose on the *Blohm & Voss Bv 141A V2*. Notice the fixed round bubble to aid in ground support activities, such as strafing and bombing.

A winter scene featuring a full head-on view of the *Blohm & Voss Bv 141A V2.* This machine appears to have highly polished aluminum propeller blades. The three-bladed propeller was spun by a *BMW 132N* 9- cylinder radial engine of 865 horsepower for up to one minute's operation. Continuous cruise horsepower was only about 664 horses.

A full port side view of the *Blohm & Voss Bv 141A V2*, *D-ORJE*. Originally known as the *Ha 141-0*, this machine with the unusual configuration was first flown on February 25th, 1938.

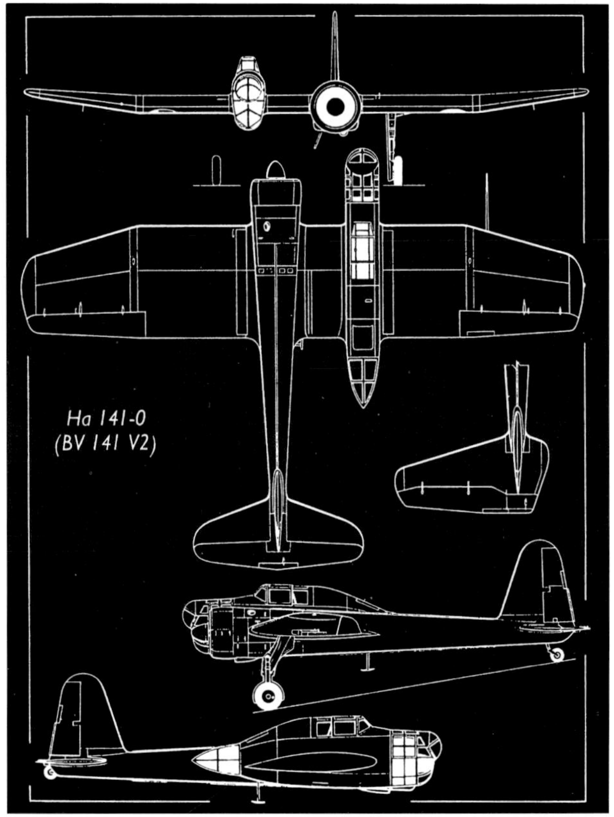

A reverse pen and ink 3-view drawing of the *Blohm & Voss Bv 141A V2.*

Ernst Udet removing his flying coat shortly after making a flight test with the *Blohm & Voss Bv 141A V2*, April 1938. Contrary to some published historical accounts, *Udet* found the machine pleasing to fly. With *Udet* are his two major aids: *Generalingneer Roluf Lucht*, to *Udet's* left and behind, and the tall *RLM* officer is *Generalingneer Gottfried Reidenbach.*

Baubeschreibung

BV141-V2

Air Documents Division, T-2
AMC, Wright Field
Microfilm No.
R 2328 F 227

BLOHM UND VOSS ABTL. FLUGZEUGBAU

aufgestellt am: 30.1.40	gesehen am: 30.1.40		Pause Nr. 3	Dr. 141/172-17/1

A poor quality copy from *Blohm & Voss* featuring the cover of their technical proposal/description of the *Bv 141A V2.* It is dated January 30th, 1940.

Blohm & Voss' preproduction prototype *Bv 141A V3 D-OLGA*. Notice that this machine appears to be painted all white except for its vertical rudder, which is black.

The *Blohm & Voss Bv 141A V3* ground level starboard nose side. It appears to still have its white underside, however, the upper surfaces have been repainted. It no longer is known as *D-OLGA*, but has radio call code *BL+AA*.

This photo shows the fuselage nose of the all white *Blohm & Voss Bv 141A V3* before it was given its radio call code *BL+AA*.

The rear cockpit cabin of the repainted *Blohm & Voss Bv 141A V3* with radio call code *BL+AA*. A rear gunner occupied this glassed over area. The cannon which would eventually be placed there was a *MG 15* 7.9 mm on a swiveling base. Range of fire up and down and from side to side. Notice the length of the landing flaps appearing on the starboard side wing.

The *Blohm & Voss Bv 141A V4* (*Ha 141 V4*) with civil registration *D-OLLE* as seen from its rear port side. The machine appears to be painted a silver color.

The *Blohm & Voss Bv 141A 01 (V4 D-OLLE)* seen after another forced landing due to hydraulic failure. This machine was the last *Bv 141A* to bear a civil registration code. It was also the first of 5 pre-series machines known as *01* to *05.* Even though they were pre-series, the *RLM* still gave them a *versuch*, or research number.

The **Blohm & Voss Bv 141A 02** (*V5* and with radio call code *BL+AB*) shown at *Erprobungsstelle*-Rechlin in July 1940.

The *Blohm & Voss BV 141A 05 (V8* with radio call code *BL+AE*) shown at *Erprobungsstelle*-Rechlin Summer 1940. When this machine's flight testing was completed it was the end of the *Bv 141A* flight test program. This was in January 1940. Afterward, all five *Bv 141A* (*V1* through *V5*) pre-series machines were delivered to various *Luftwaffe* units for use under field conditions and evaluation. It seems like a useless effort, because the *OKL* had decided by April 4th, 1940, that the *Bv 141A* would not be placed in series production.

A starboard side view of *Blohm & Voss Bv 141A V8's* tail assembly. Pretty much a standard configuration. However, *Richard Vogt* would change all this in the re-engineered *Bv 141B*...making its tail asymmetrical.

The staboard side of a *Blohm & Voss Bv 141B-01* (*V9*) flying low over the German countryside with rivers. The *V9* was an entirely redesigned machine and equipped with a *BMW 801N* air cooled radial 14-cylinder engine of 1,665 takeoff horsepower. *Richard Vogt* was given permission from the *RLM* to re-engine his *Bv 141* in January 1939. It would become virtually a new airplane. *Vogt* had a mockup ready to inspect on February 14th, 1940. Its maiden flight was made on January 9th, 1941.

Left: A close-up view of *Blohm & Voss Bv 141B-01 (V9)* with more of the German farmland in the background.

Opposite: A 3-view drawing by *Blohm & Voss* with performance data of their *Bv 141B* with the *BMW 801N* radial engine as presented to the *RLM*.

R L M	Flugzeugtypenblatt	Baumuster: BV 141-B

Bildskizze: Maßstab: 1:200

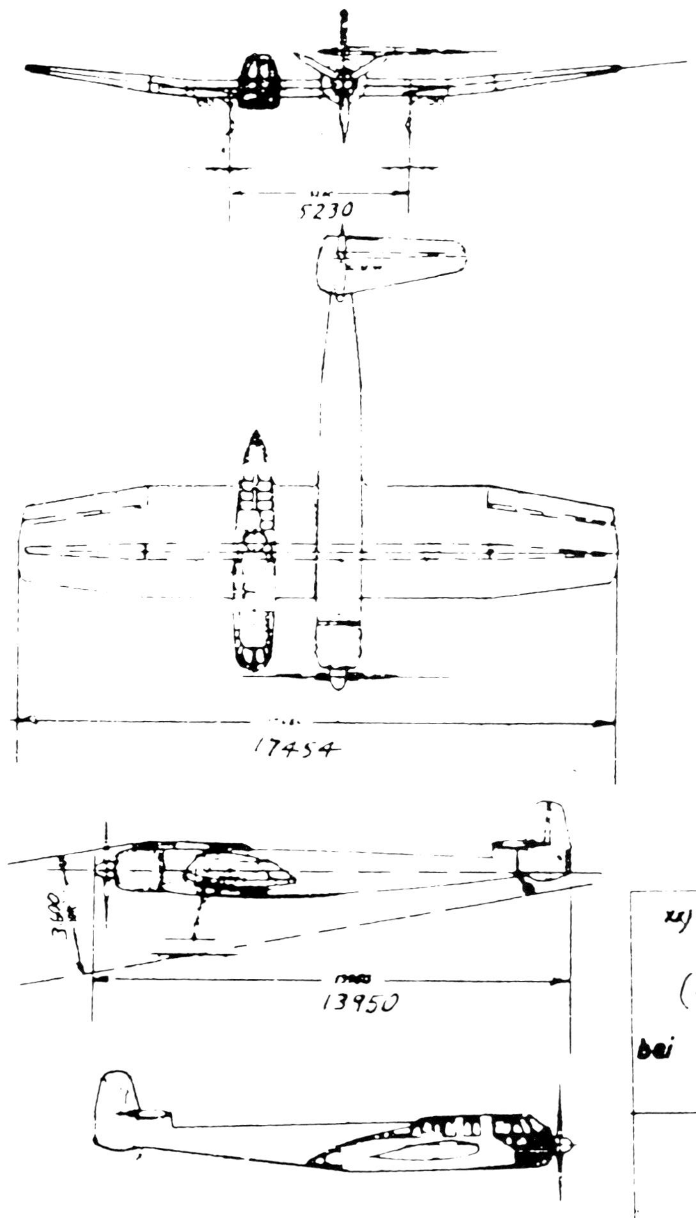

xx) Flugweite in H=5 km bei Sparleistung (n_M = 2000 U/Min) bei B_K = 750 kg / B_K = 480 kg

* Motor Luftschraube ~~Laufrädern~~

Bauart: B | Bauweise: Metall | Verwendungszweck Nahaufklärer

Motortype	BMW 801	Luftschrauben Type		Zuladungs-Gewicht:	1360 kg	Höchst-geschwindigk.	405 km/h (H=0) 440 km/h (H=4,1)
Motoranzahl	1	Luftschraube-Gewicht	~ 150 kg	Flug-gewicht:	5660 kg	Flugweite: xx)	1600 km 1040 km
Motorgewicht	1120 kg	Leergewicht (ohne) *	3030 kg	Fläche:	52,9 m²	Dienst-gipfelhöhe	8900 m
Höchst leistung	N_1' = 1560 PS H = 0 km	Leergewicht (mit) *	4300 kg	Flächen-belastung	107 kg/m²	Leistungs belastung	$\frac{G}{N_1'}$ = 3,63 kg/PS
Drehzahl pro Min	n_M = 2700 U/Min bei N_1'	Rüstgewicht	4300 kg	Reise geschw	425 km/h bei 385 km/h bei	$N_{Dauer\,max}$ N_{Spar}	(2300 U/Min) (2000 U/Min) H=5 km

A poor quality photo of *Blohm & Voss Bv 141B-01 V9 (NC+OZ)* in low level flight.

The *Blohm & Voss Bv 141B-02 V10* with radio call code *NC+RA* seen shortly after takeoff.

A rare picture indeed. Three *Blohm & Voss Bv 141Bs—V9, V10,* and *V11*—are shown lined up at *Erprobungsstelle*-Rechlin early Summer 1941.

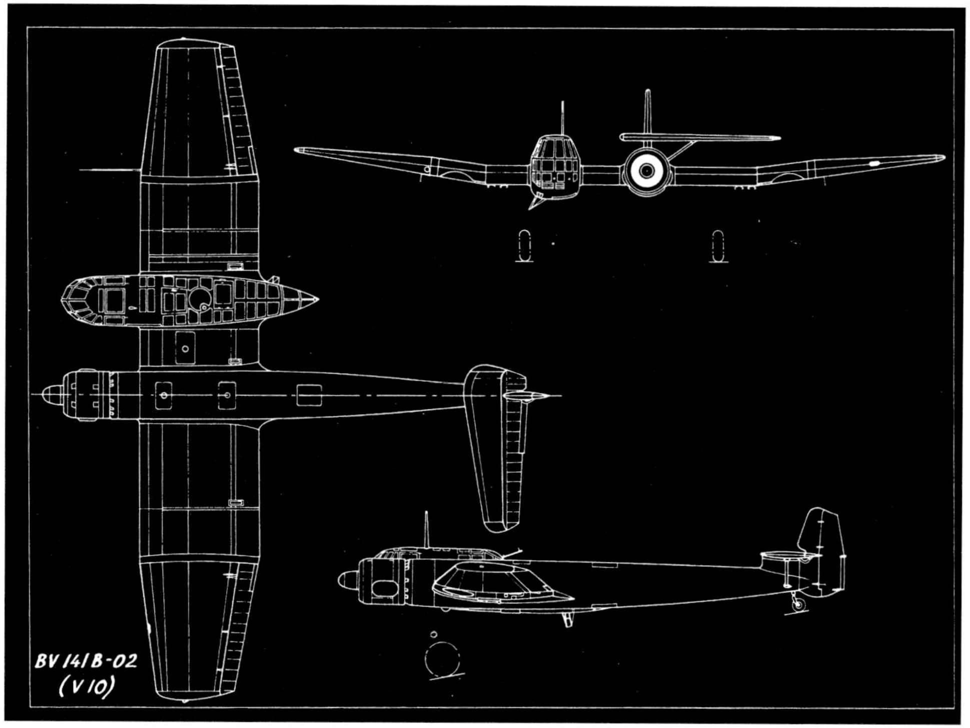

A pen and ink 3-view reverse negative of the *Blohm & Voss Bv 141B-02 V10.*

The *Blohm & Voss Bv 141B-02 V10* and radio call coded *NC+RA.* Shown at *Erprobungsstelle* Rechlin Summer 1941. It was later transferred to *Aufklärungsschule 1*, Grossenhain, Autumn 1941 for field service trials. Its final disposition is unknown.

One of the nicest photos of *Richard Vogt's Blohm & Voss Bv 141* prototype which survived the war. This machine, as seen from its starboard side, is *Bv 141B-03, V10, werk numer 100003*, with radio call code *NC+RB*. It was powered by a *BMW 801N* 14-cylinder "double star."

An overhead starboard side view of the *Blohm & Voss Bv 141B V11*, which appears being readied for flight.

Blohm & Voss Bv 141B-04 (*Bv 141 V12*) with radio call code *NC+RF* seen taking off from a snow-covered runway somewhere in Germany.

A mystery *Blohm & Voss Bv 141B*. This *Bv 141B* was found on a remote *Luftwaffe* air base under some camouflage netting as seen from its starboard side. All of its cockpit cabin windows appear to be knocked out. Also, its tail assembly has been severed from its fuselage. There is a tail assembly lying on the ground behind the machine, however, it is not the tail assembly which belonged to the *Bv 141B*. This is because the tail assembly lying there on the ground is from a *Bv 141A*! At some point in time, the *Bv 141A* tail assembly must have been dumped right behind the fuselage of the *Bv 141B*. But what is this machine with the radio call code (*G*)*K+GH*? Where does it belong in the sequence of the 20 *Bv 141s* which *Richard Vogt* claimed, postwar, were constructed by *Blohm & Voss?*

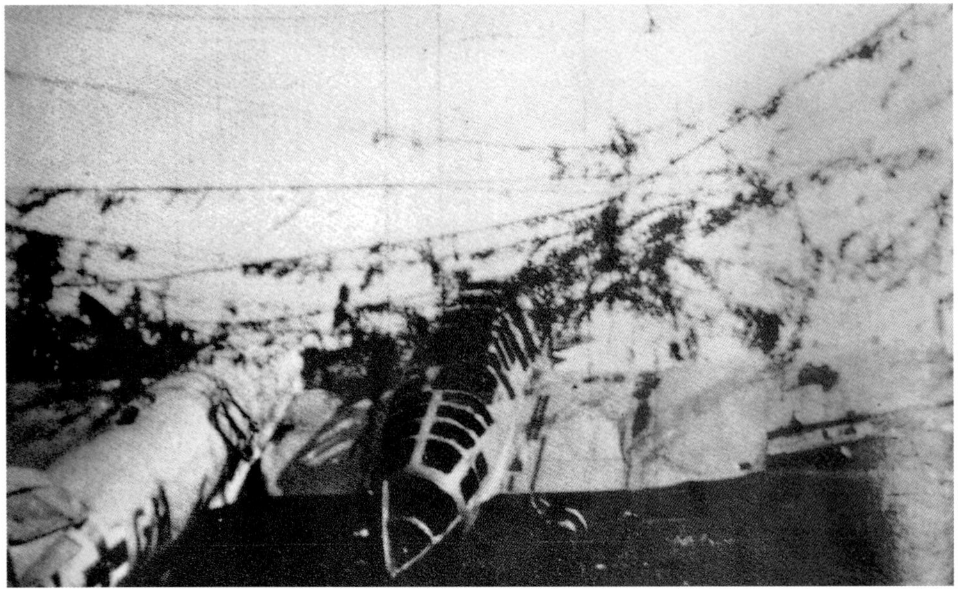

A poor quality photo of *Blohm & Voss Bv 141B* with the radio call code *(G)K+GH* taken from behind the cockpit cabin.

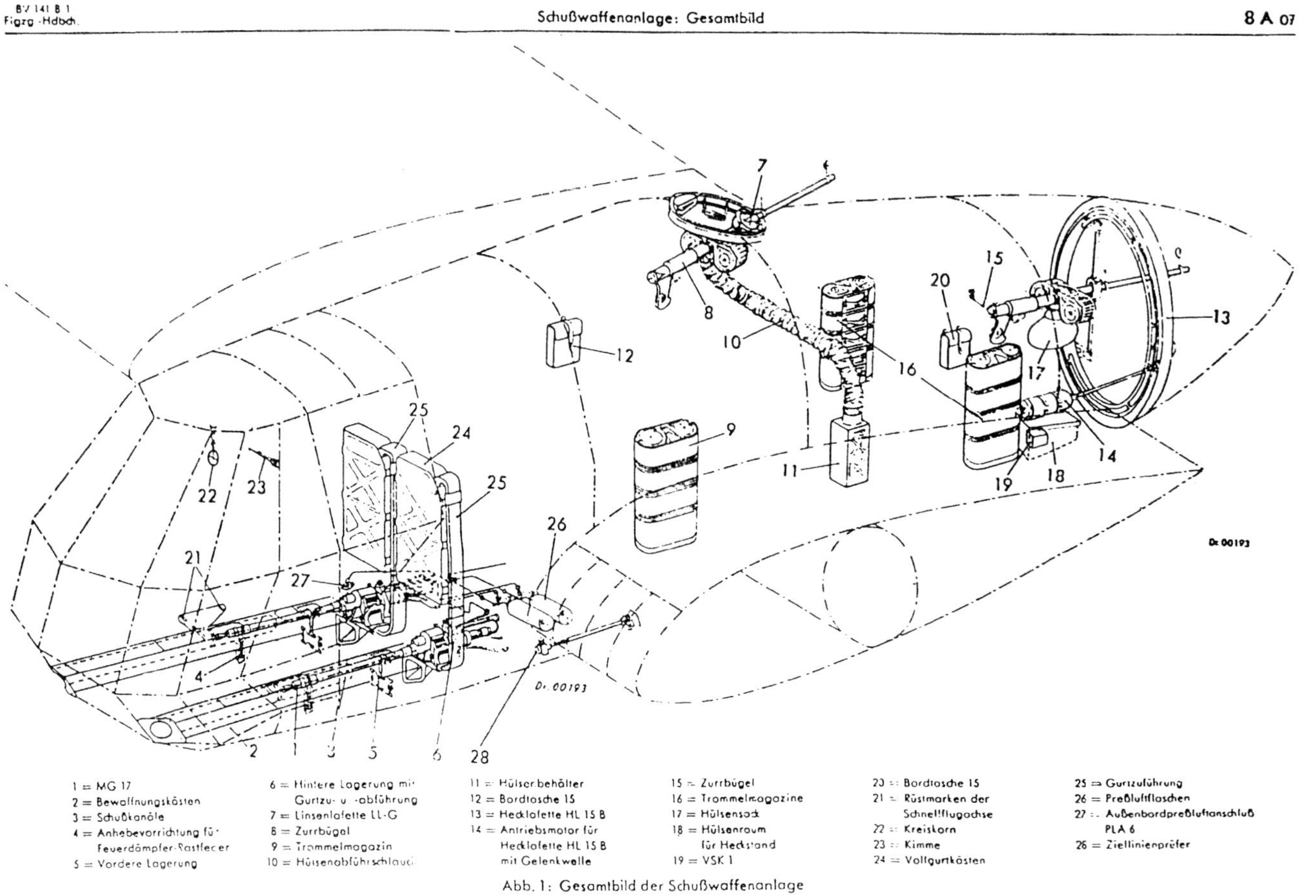

A pen and ink illustration from *Blohm & Voss* and featuring the *Bv 141B's* overall defensive armament system as seen from its port side. Each *Bv 141B* series would have had four *7.9 mm* cannon: two fixed *MG 17* in the cockpit cabin's nose, a single movable *MG 15 7.9 mm* in a dorsal turret, and another single swiveling *MG 15 7.9 mm* cannon in the apex of the cockpit cabin similar to that found in the *Fw 189A*:

01 - *MG 17* cannon
02 - weapon container
03 -
04 -
05 - front storage room
06 - back storage room featuring belt-feeding and discharge (spent cartridge) mechanism
07 - lens-shaped machine gun mounting
08 -
09 - drum magazine
10 - spent cartridge hose
11 - cartridge container
12 - on board storage sack
13 - rear cannon mounting
14 - drive motor for the rear cannon with drive shaft
15 -
16 - drum magazine
17 - cartridge sack
18 - cannon cartridge space for rear gun turret
19 -
20 - on board storage sack
21 -
22 -
23 -
24 - cartridge-belt container
25 - cartridge-belt feeding mechanism
26 - air pressure bottles for the PLA-6
27 -
28 -

A ground level nose on view of the *Blohm & Voss Bv 141B's* cockpit cabin featuring the location of its twin fixed *MG 17 7.9 mm* cannon. They were installed off to port so to allow an observer space to see out and below. Notice that one *MG 17* cannon port is far to the port side of the cabin, while the other cannon port is about on the cabin's center line.

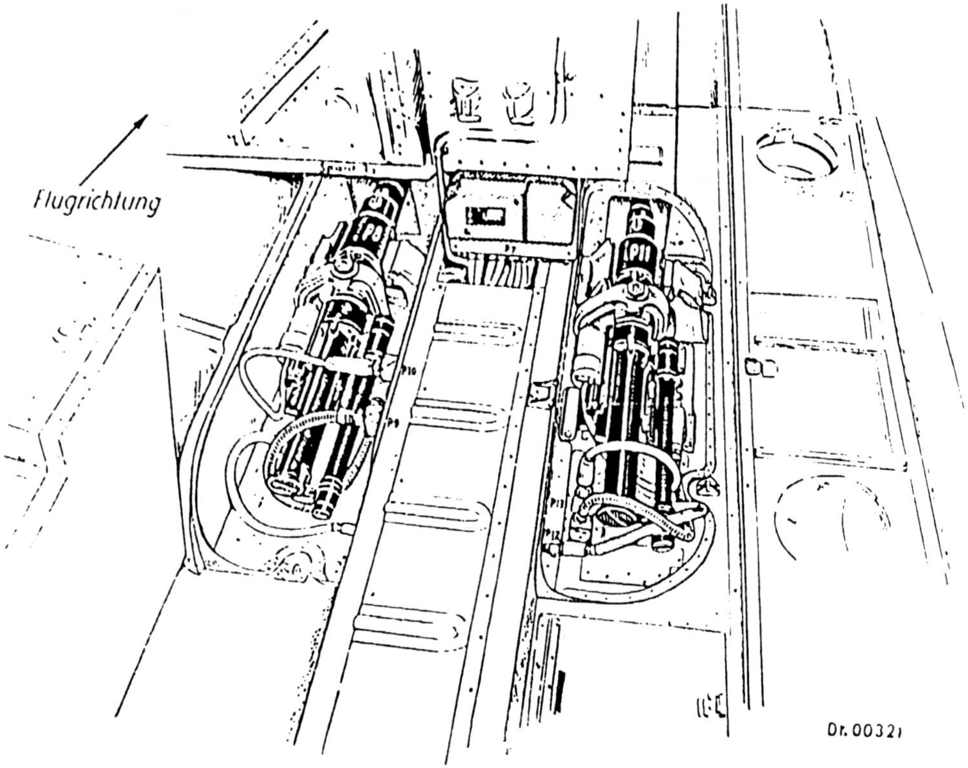

Abb. 26: Starre Schußwaffe

A pen and ink illustration from *Blohm & Voss* featuring the mounting position of its two fixed *MG 17 7.9 mm* cannon in the *Bv 141B's* cockpit cabin's floor. These cannon are shown without their ammunition storage and feed mechanism.

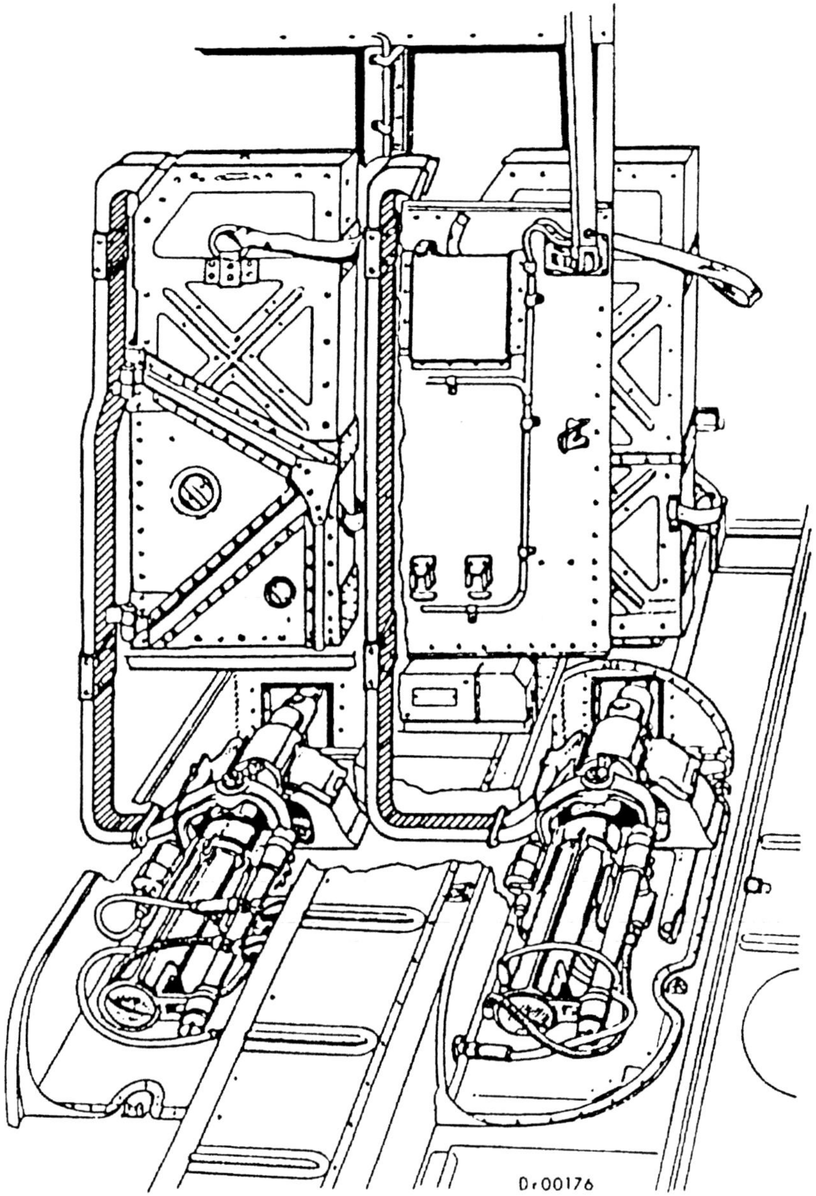

A pen and ink illustration from *Blohm & Voss* featuring the cartridge belt-feeding plan fixed 2x*MG 17 7.9 mm* nose-mounted cannon on the *Bv 141B*, including the ammunition container for both cannon.

Abb. 10: Gurtzuführung der MG 17

A pen and ink sketch featuring the starboard side of a *Blohm & Voss Bv 141B*. Notice the approximate position on the cockpit cabin where its rear-firing swivelling *MG 15 7.9 mm* cannon is located.

An overhead view of the *Blohm & Voss Bv 141B's* cockpit cabin. Notice the egg-shaped rear gunner's canopy and how the single swivelling *MG 15 7.9 mm* cannon is off to port. Aft and down is the *Bv 141's* second cockpit cabin hatch. The other is located directly forward the *MG 15 7.9 mm* gunner's oval canopy. Courtesy: *Gert W. Neumann.*

A stunning view of *Blohm & Voss Bv 141B V9* and its cockpit cabin as seen from its rear starboard side. The forward hatch is open, and immediately aft is its single swivelling *MG 15 7.9 mm* cannon turret. No cannon appears in the photograph.

A view of *Blohm & Voss Bv 141B V9's* aft cockpit cabin with both entry/exit hatches open. Notice that they open to port.

A view of the *Focke-Wulf Fw 189A's* port side cockpit cabin. Note the amazing similarity between the two machines. However, the *Blohm & Voss Bv 141B*'s cannon protruded out an open lower panel. The *Fw 189A's* cannon opening was an upper panel.

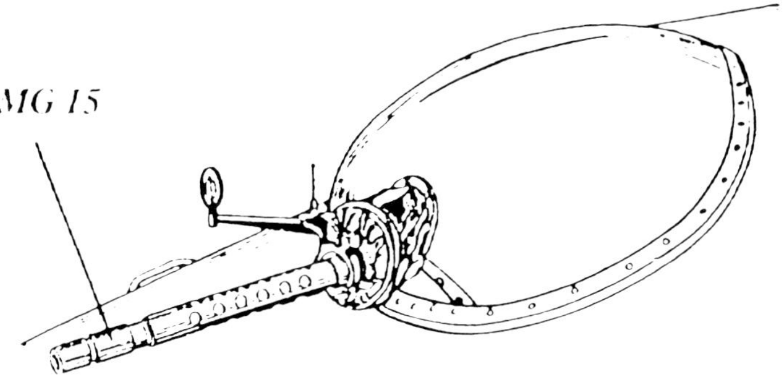

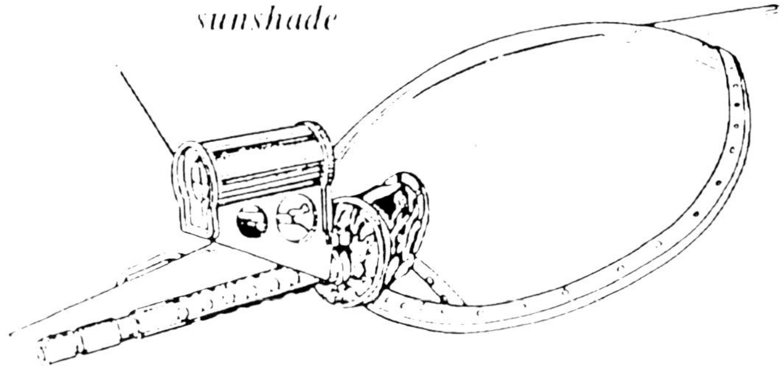

A pen and ink illustration featuring the *MG 15 7.9 mm* fitted in a swivelling turret. The bottom image shows the *MG 15* fitted with a sunshade.

A gunner with a single swivelling *MG 15 7.9 mm* cannon, similar to that found on the *Blohm & Voss Bv 141B.*

A seris of poor quality photos featuring the *Blohm & Voss Bv 141's* aft cabin and swivelling *MG 15* 7.9 mm cannon turret.

Top Left: The *MG 15* cannon's barrel is seen positioned far to port and elevated as high as it is able.

Top Right: The *MG 15* cannon positioned in its resting position...center and straight ahead.

Bottom Right: The *MG 15* cannon as seen from outside the *Bv 141* in its open bay.

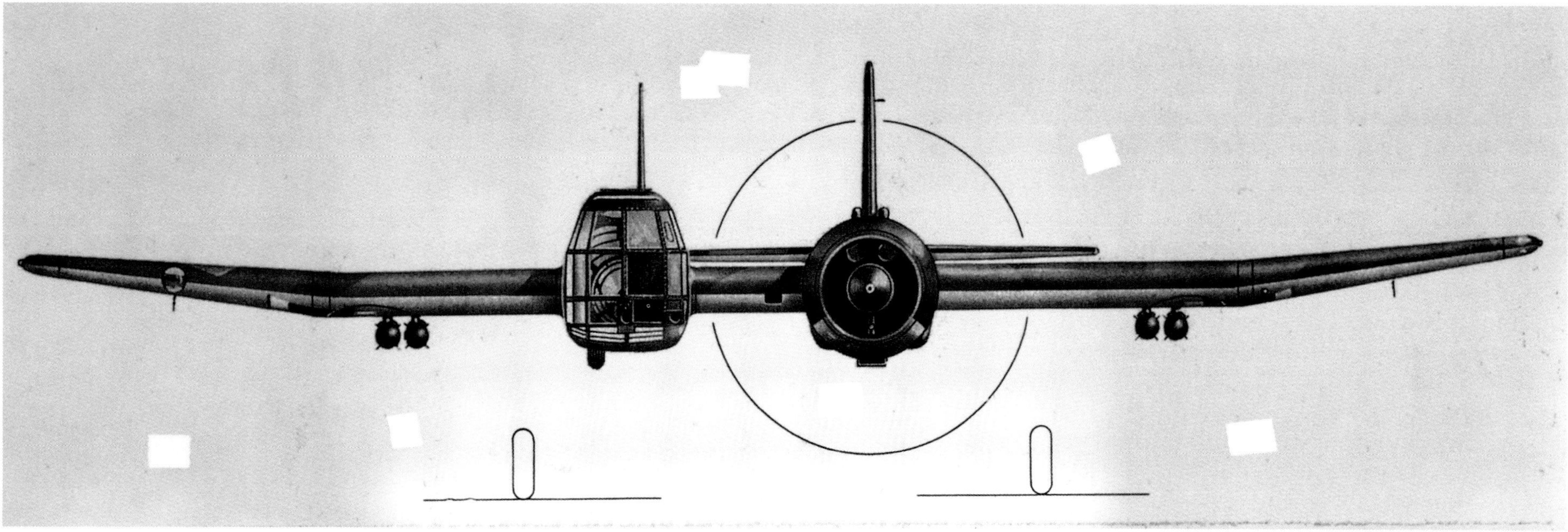

A ground level nose on view of a *Blohm & Voss Bv 141B*. Under each wing are two *ETC 50 VIII* bomb racks. Courtesy: *Gert W. Neumann.*

A poor quality port side wing view of a *Blohm & Voss Bv 141B* and its low profile *ETC 50* bomb rack. Notice that the *ETC* rack is located aft the main wheel (gear) well.

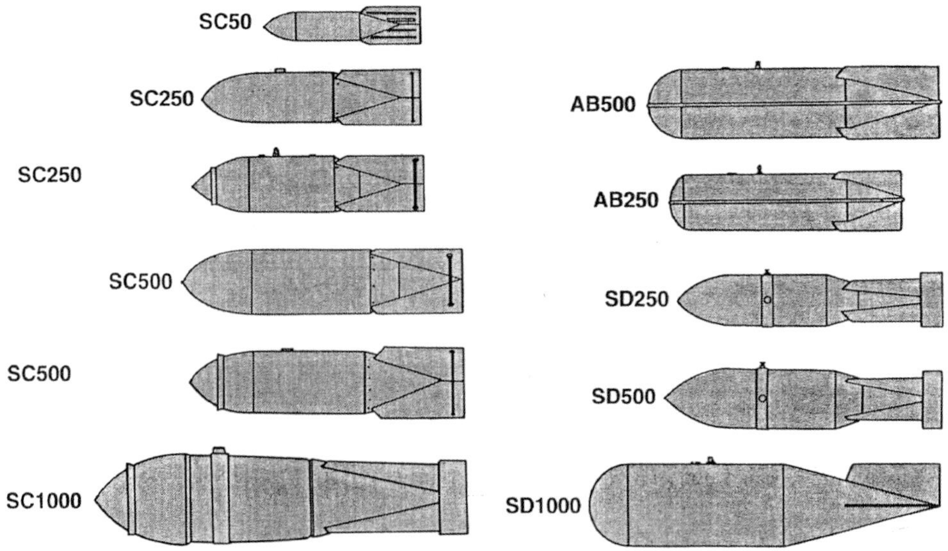

The *Blohm & Voss Bv 141B* normally would have been fitted to carry 4x*SC50* (4x110 pounds) bombs. The lightest of the *Luftwaffe* inventory. On occasion it could be fitted to carry 1x*SC250* (551 pounds).

For *Wehrmacht* ground support, the *Blohm & Voss Bv 141B* could carry 4x*SC50* (4x110 pounds) of the type shown in this photograph. These *SC50s* have been fitted with stand-off pipes...the so-called "*Dinort Spargeln*," or *Dinort asparagus*. These weapons exploded just above the ground and were very effective against vehicle convoys and troop assembly areas. The *SC50* was named after *Oskar Dinort*, their inventor, a noted *Stuka* pilot and commander. *Oskar Dinort* also befriended the *Horten* brothers in the early 1940s, and had *Reimar Horten* build his all-wing *Ho 3* sailplanes for the *Dinort* competition team at the *Rhön-Wasserkupper*.

Oskar Dinot.

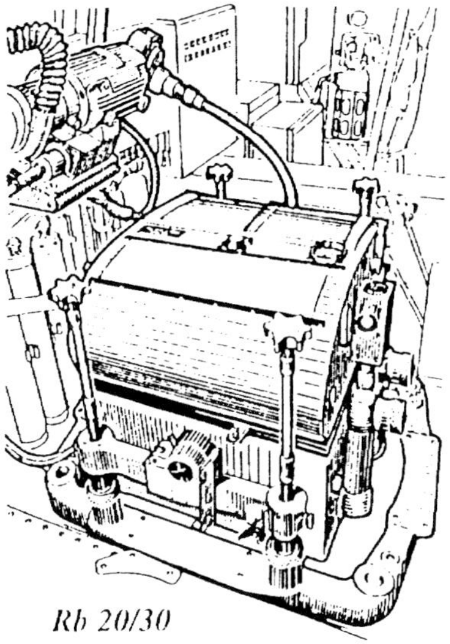

Rb 20/30

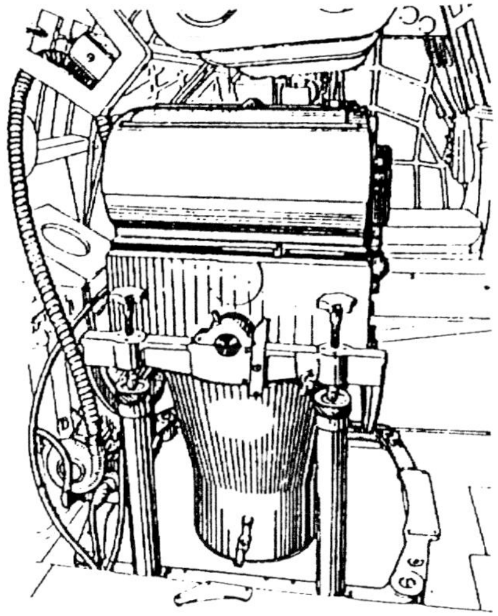

Rb 50/30

The *Blohm & Voss Bv 141B* was designed to carry one automatic reconnaissance camera, either an *Rb 20/30, Rb 50/30,* or an *Rb 21/18.* The two most common cameras to be used in the *Bv 141B* would have been the *Rb 20/30* or *Rb 50/30*, as featured in these pen and ink illustrations.

The port side main landing gear of a *Blohm & Voss Bv 141A* undergoing statics tests. The main wheel hydraulics was a source of constant problems on the *"A"* version.

The hydraulic retraction cylinder seen on a *Blohm & Voss Bv 141A's* port side.

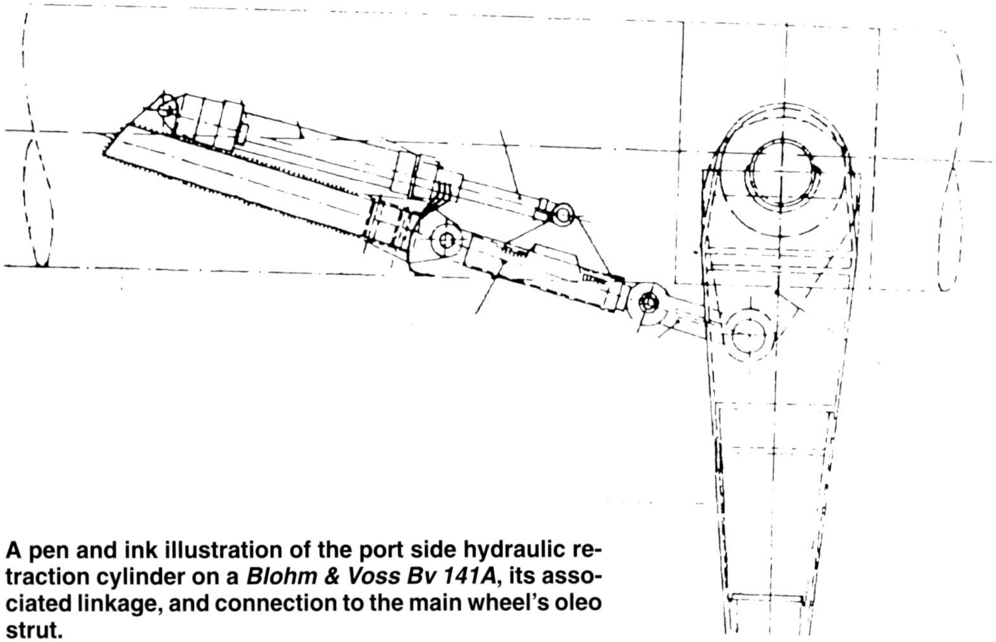

A pen and ink illustration of the port side hydraulic retraction cylinder on a *Blohm & Voss Bv 141A*, its associated linkage, and connection to the main wheel's oleo strut.

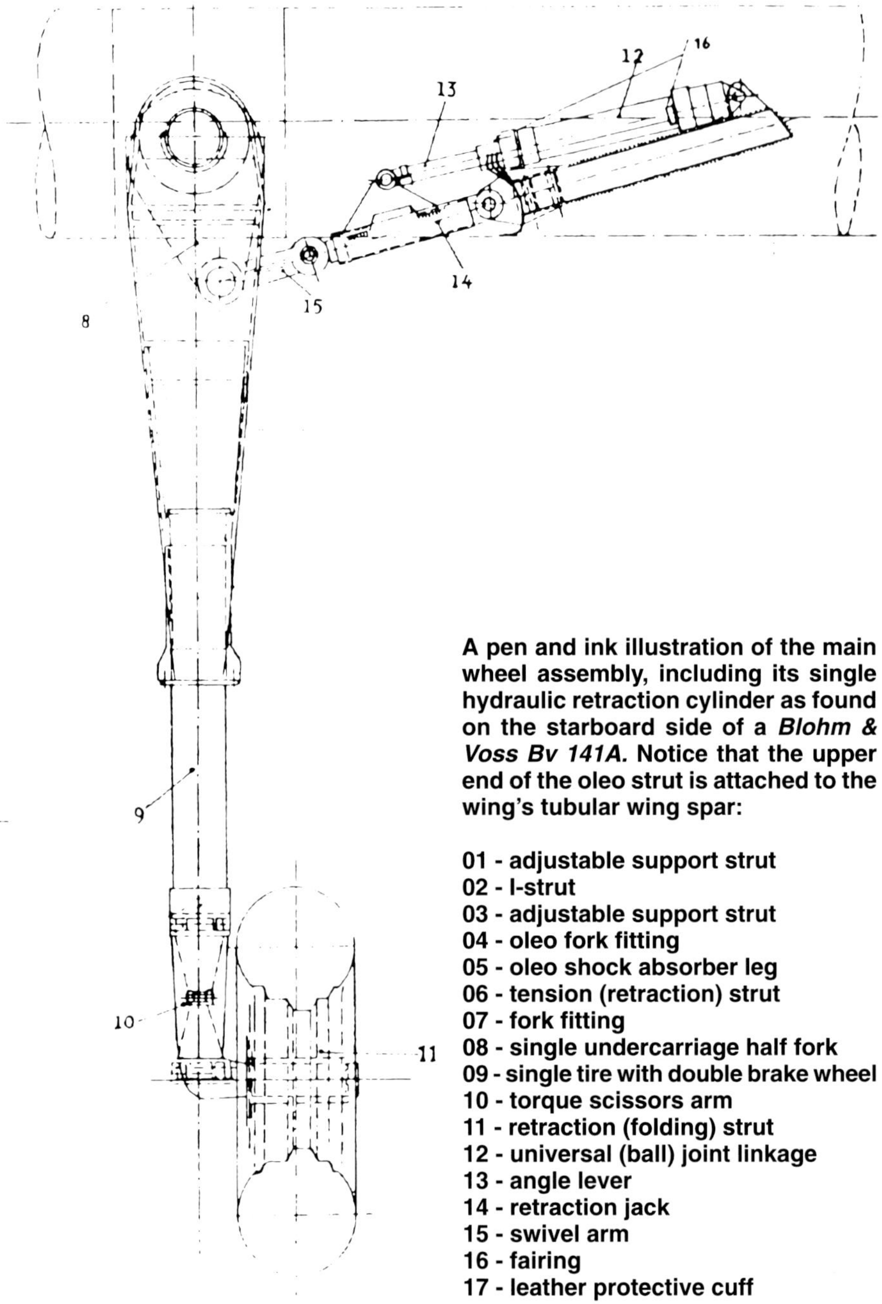

A pen and ink illustration of the main wheel assembly, including its single hydraulic retraction cylinder as found on the starboard side of a *Blohm & Voss Bv 141A.* Notice that the upper end of the oleo strut is attached to the wing's tubular wing spar:

01 - adjustable support strut
02 - I-strut
03 - adjustable support strut
04 - oleo fork fitting
05 - oleo shock absorber leg
06 - tension (retraction) strut
07 - fork fitting
08 - single undercarriage half fork
09 - single tire with double brake wheel
10 - torque scissors arm
11 - retraction (folding) strut
12 - universal (ball) joint linkage
13 - angle lever
14 - retraction jack
15 - swivel arm
16 - fairing
17 - leather protective cuff

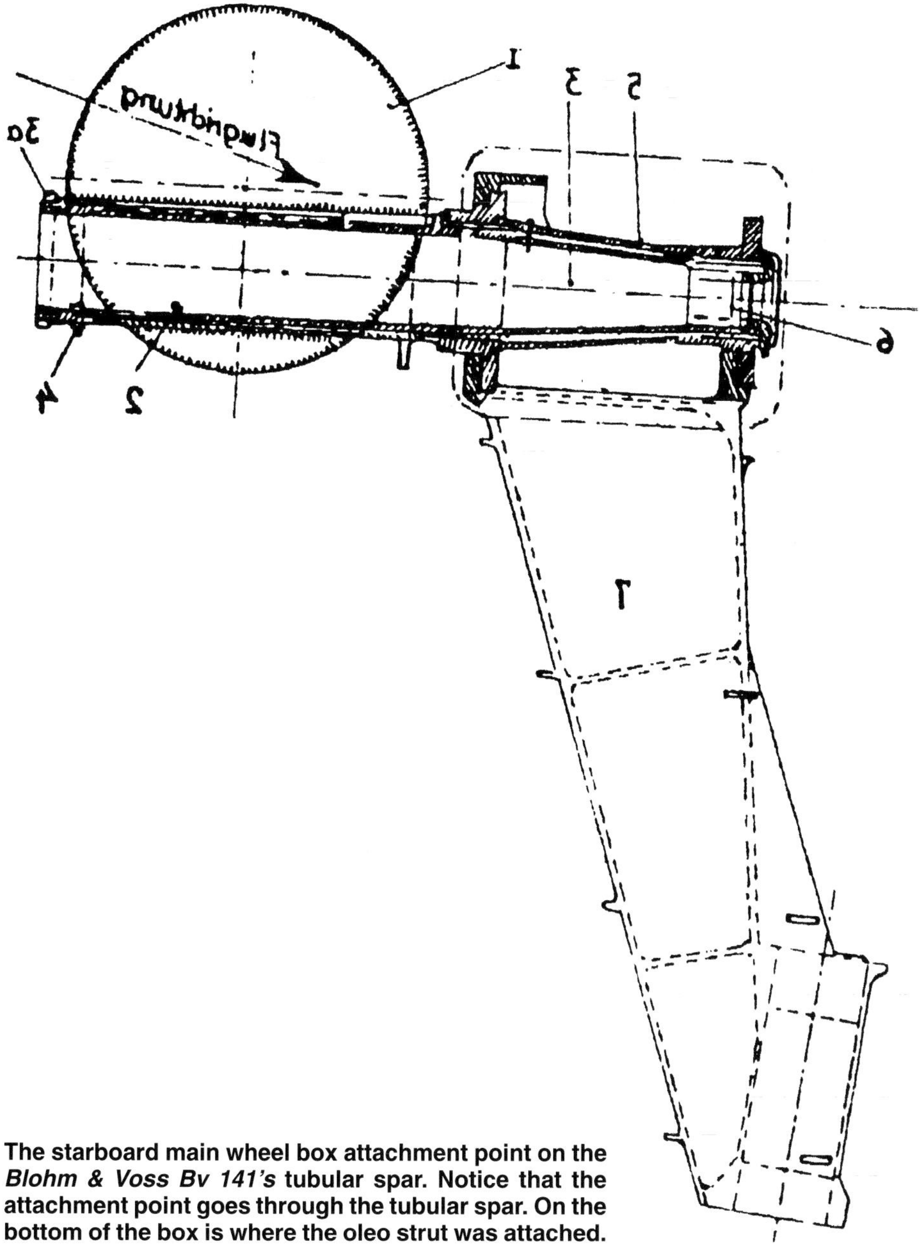

The starboard main wheel box attachment point on the *Blohm & Voss Bv 141's* tubular spar. Notice that the attachment point goes through the tubular spar. On the bottom of the box is where the oleo strut was attached.

The starboard box/oleo strut on a *Blohm & Voss Bv 141A* can be clearly seen in this photo. Notice that the crew member is taking advantage of the built-in foot support in climbing up and into the cockpit cabin.

A poor quality photo of the starboard side boxed main landing gear assembly on a *Blohm & Voss Bv 141A*. Also, dimly shown, is the starboard side gear assembly's storage area just aft the wing's leading edge.

When the *Blohm & Voss Bv 141A* was re-engined, thus becoming the *Bv 141B*, *Richard Vogt* re-designed the landing gear, too. The older style *"L"* shaped ole strut became a half fork style.

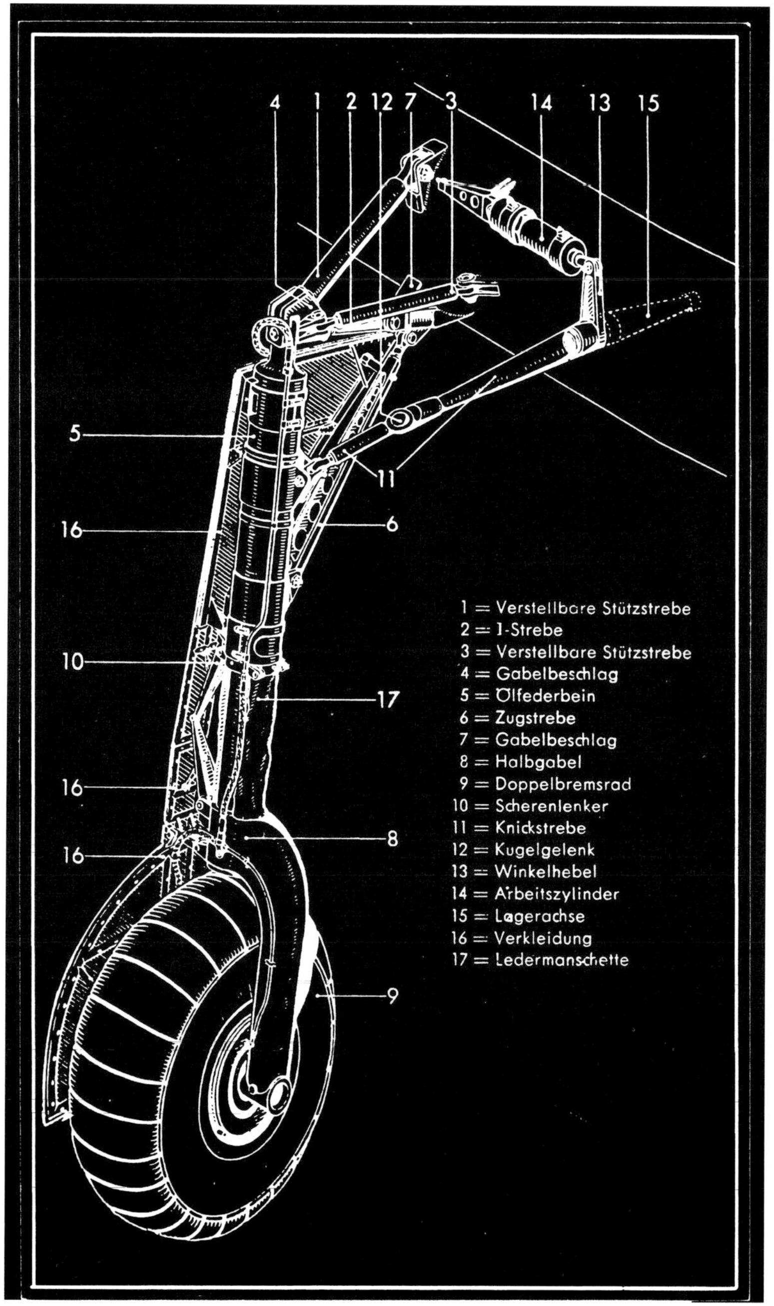

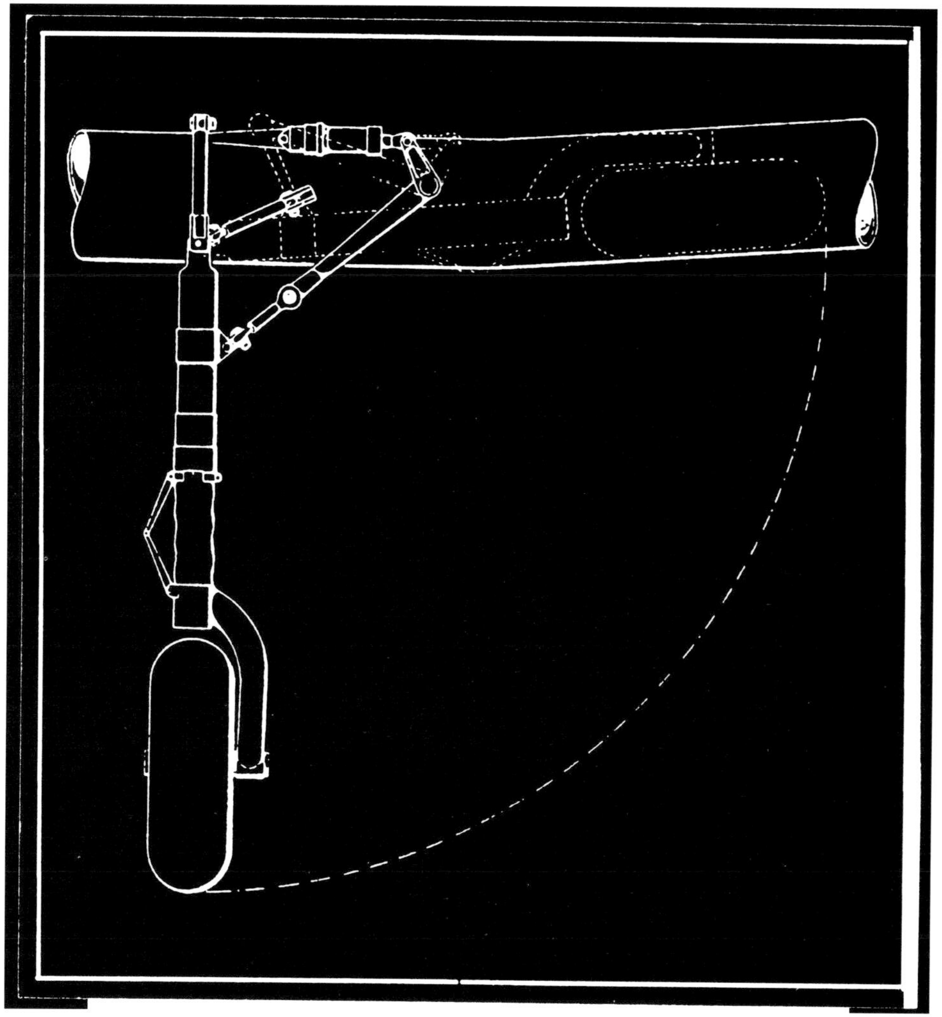

A pen and ink reverse negative illustration of the port side main landing gear retraction movement on a *Blohm & Voss Bv 141B* and its storage box within the port wing. Drawing by *Günter Sengfelder*:
01 - wheel fairing
02 - screw cap
03 - towing lug

A pen and ink reverse negative illustration of the re-engineered main wheel landing gear on the *Blohm & Voss Bv 141B*. This drawing represents the port side landing gear assembly. Drawing by *Günter Sengfelder*.

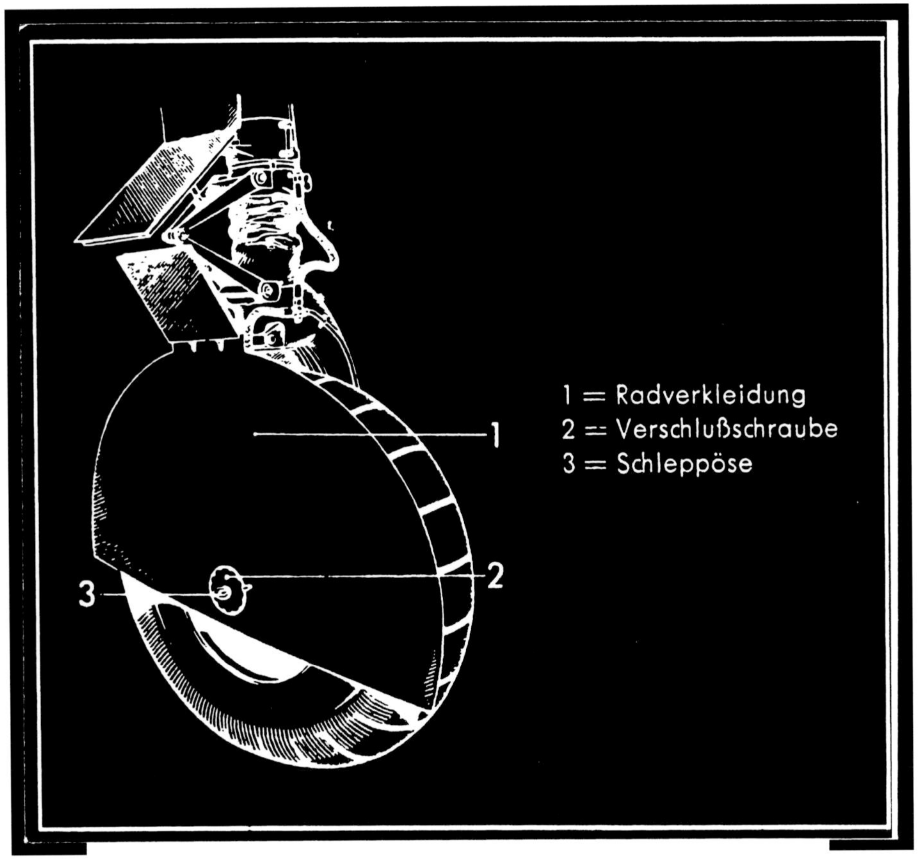

A pen and ink reverse negative illustration featuring the port side main landing wheel, its cover, and scissor-fork on the *Blohm & Voss Bv 141*. Drawing by *Günter Sengfelder.*

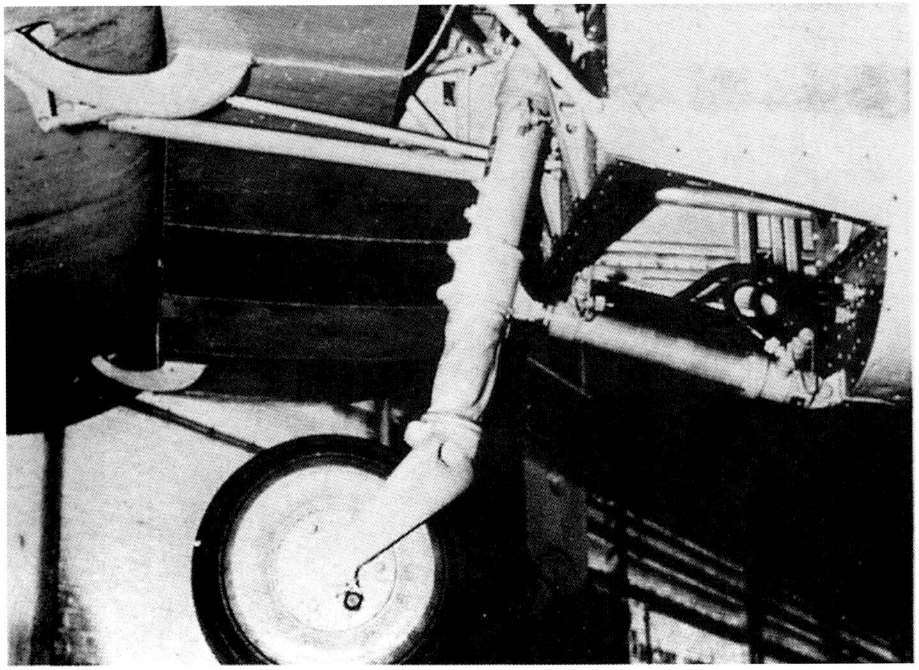

The starboard side tail wheel assembly on the *Blohm & Voss Bv 141A.*

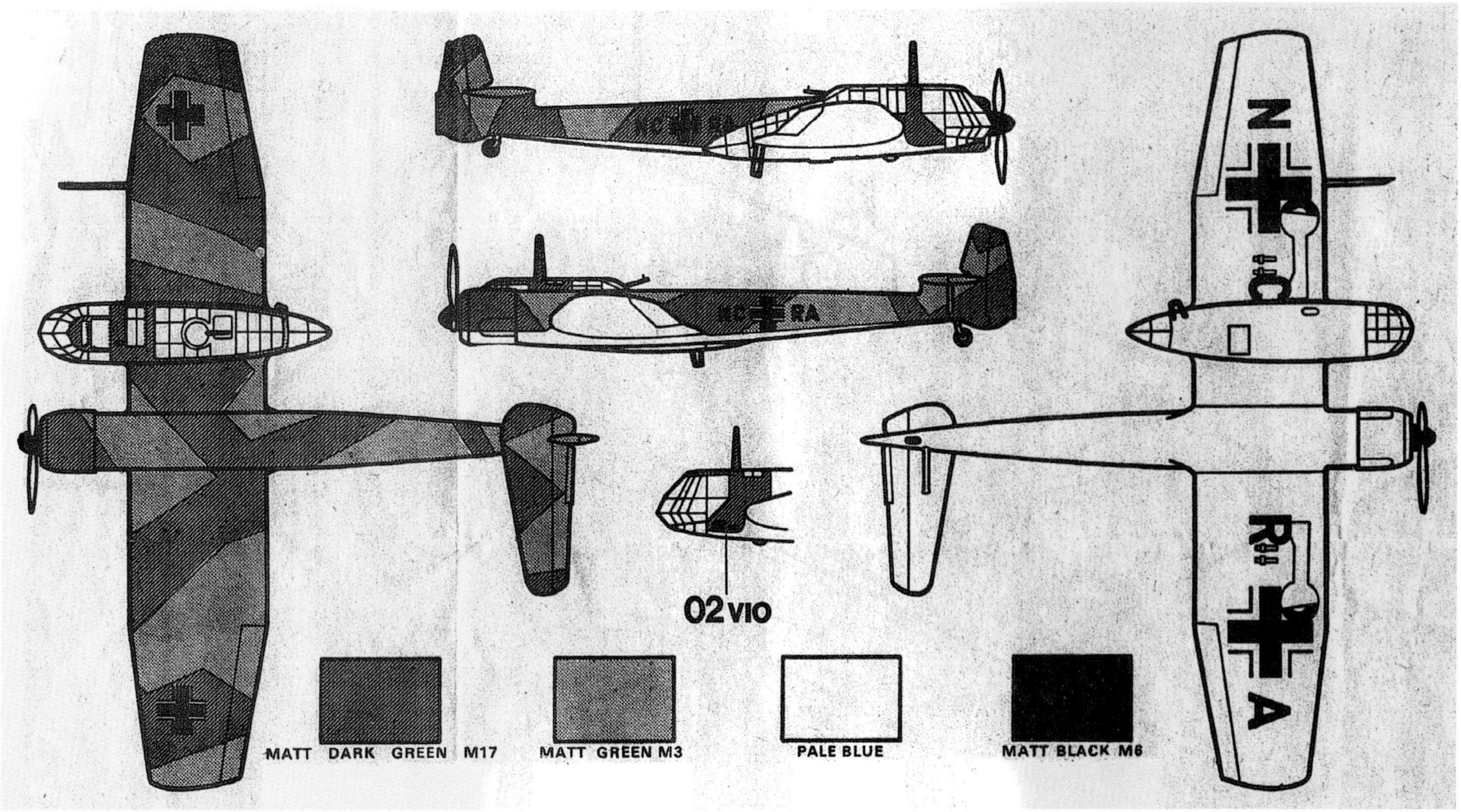

Typical camouflage color/patterns on the *Blohm & Voss Bv 141B*. Courtesy: *Airfax* 1/72 scale model construction kit.

4 34 Steuerwerk: Handlochdeckel- und Klappenübersicht BV 141 B-1 Flgzg.-Hdbch.

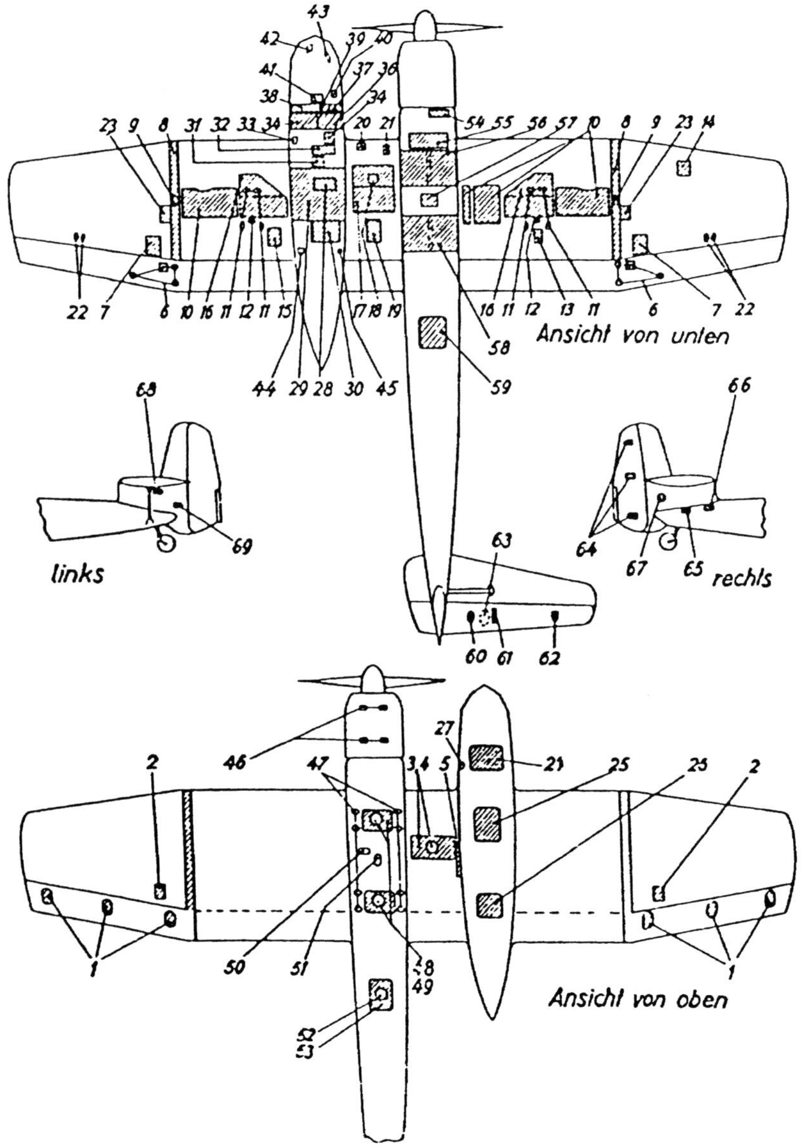

Zeichnung 8: Handlochdeckel- und Klappenübersicht

Manual access service covers and panels:

01 - lower wing surface cover for the aileron adjustment control
02 - lower wing surface cover for the aileron control rod to aileron
03 - lubrication cover
04 - fuel filter cover
05 - hoisting eye cover
06 - upper wing aileron trim adjustment cover
07 - upper wing aileron trim control adjustment cover
08 - slot cover
09 - cover slot
10 - rib attachment cover
11 - ETC cover
12 - ETC covering
13 - drain cover
14 - landing light cover
15 - landing flap activation cylinder cover
16 - swing panel for spar and undercarriage
17 - double panel for lubricating landing control mechanism
18 - panel for lubrication drain port
19 - panel for accessory equipment
20 - panel
21 - panel for electric equipment
22 - access panel
23 - aileron trim access panel
24 - removable emergency window
25 - forward entry hatch
26 - aft entry hatch
27 - electric access panel

A *Blohm & Voss* logo featuring their asymmetrical *Bv 141B.*

A magazine advertisement by *BMW* featuring the asymmetrical *Blohm & Voss Bv 141B,* and, of course, powered by a *BMW* engine.

Flugwerk für Flugzeuge mit Kraftantrieb (Gr. 3—24).

b 3_{02} Pat. 685 480 v. 16. 5. 35, veröff. 18. 12. 39. **Blohm & Voß und Dr.-Ing. Richard Vogt, Hamburg.** *Flugzeug mit einem oder mehreren Motoren.*

Patentansprüche:

1. Flugzeug mit einem Motor oder mehreren Motoren und einer oder mehreren Luftschrauben, die im letzten Fall im gleichen Luftstrahl liegen, dadurch gekennzeichnet, daß bei unsymmetrischem Aufbau das Triebwerk einerseits und ein triebwerkloser Rumpf oder eine triebwerklose Gondel andererseits aus der Mitte gesetzt sind.

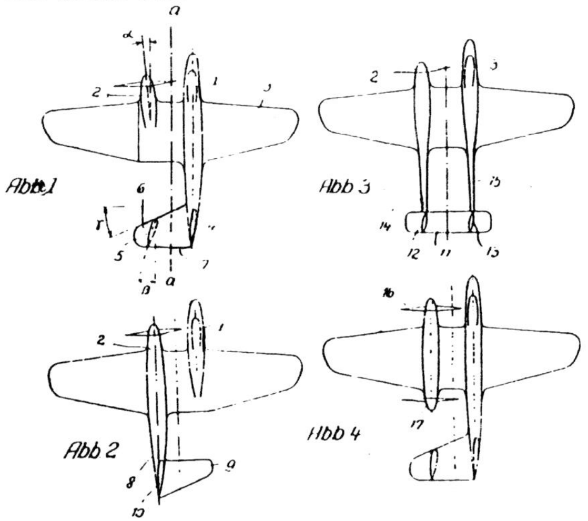

2. Flugzeug nach Anspruch 1, dadurch gekennzeichnet, daß die Luftschraubenzugrichtung einen vom Rumpf (1) abstrebenden Winkel α mit der Längsachse (α-α) bildet.

3. Flugzeug nach Anspruch 1, dadurch gekennzeichnet, daß der im Schraubenstrahl liegende Teil des Seitenleitwerkes (5) einen Winkel β mit der Längsachse (α-α) bildet.

4. Flugzeug nach Anspruch 1, dadurch gekennzeichnet, daß die Vorderkante (6) des Höhenleitwerkes (7) mit der Propellerachse bzw. mit der Längsachse (α) einen Winkel bildet.

Left: *Richard Vogt* applied for numerous patents for his asymmetrical planform. This is one from December 18th, 1939. Notice the four pen and ink drawings featuring single engine aircraft with asymmetrical planforms accompanying the patent statement

b 43_{04} Pat. 695 920 v. 2. 7. 38, veröff. 5. 9. 40. **Blohm & Voß und Dr.-Ing. Richard Vogt, Hamburg*).** *Flugzeug mit mehreren Schwimmkörpern.*

Man kann den Luftwiderstand bei Flugbooten vermindern, wenn man die Schwimmwerksanordnung erfindungsgemäß unsymmetrisch ausführt, d. h. eine unsymmetrische Anordnung zweier Schwimmkörper wählt. Unter Schwimmkörpern sind sowohl Nutzlast tragende (Flugboote) wie auch Nur-Schwimmer zu verstehen. So ergibt sich eine der Erfindung gemäße Anordnung.

Die Schwerpunktssymmetrielinien sind strichpunktiert und mit A—B bezeichnet.

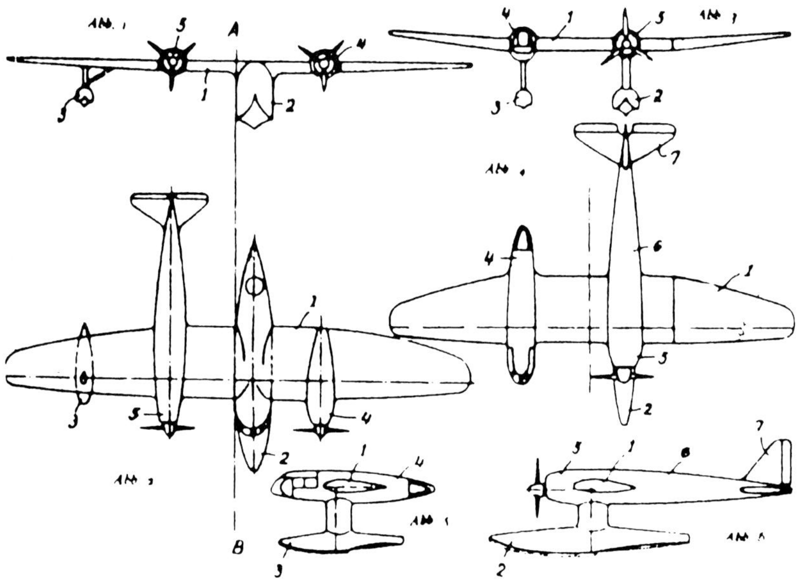

Patentanspruch:

Flugzeug mit mehreren Schwimmkörpern, gekennzeichnet durch unsymmetrische Anordnung zweier Schwimmkörper.

*) Dr.-Ing. Richard Vogt, Hamburg, ist als Erfinder genannt worden.

Right: A *Richard Vogt* patent statement from September 5th, 1940, and featuring two seaplane designs with asymmetrical planforms.

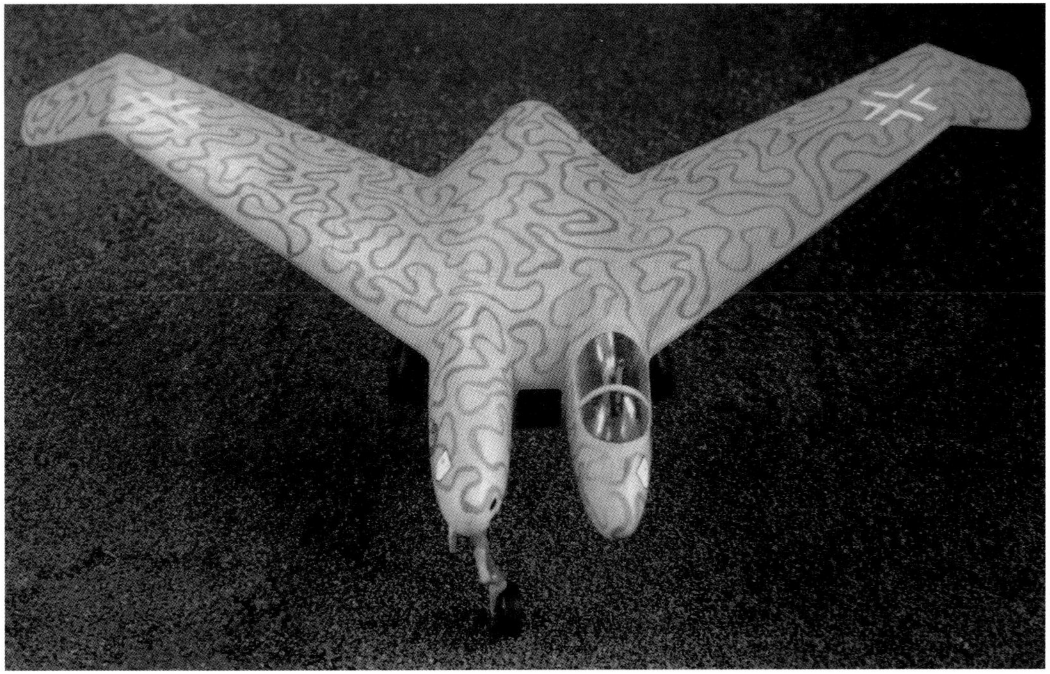

Richard Vogt was not the only German aircraft designer to see the benefits of an asymmetrical aircraft design. This is the *Heinkel P.1078B,* a single seat tailless day fighter powered by 1x*HeS 011A* turbojet engine of 2,866 pounds [1,300 kilograms] thrust from Heinkel-Hirth. Its design was proposed by *Siegfried Günter* near war's end, and it was not pleasing at all to *Ernst Heinkel. Günter* reasoned that the machine's armament, electronics, and so on could all be housed conveniently in the starboard nacelle. The pilot sat comfortably in the port nacelle. Scale model and photographed by Dan Johnson.

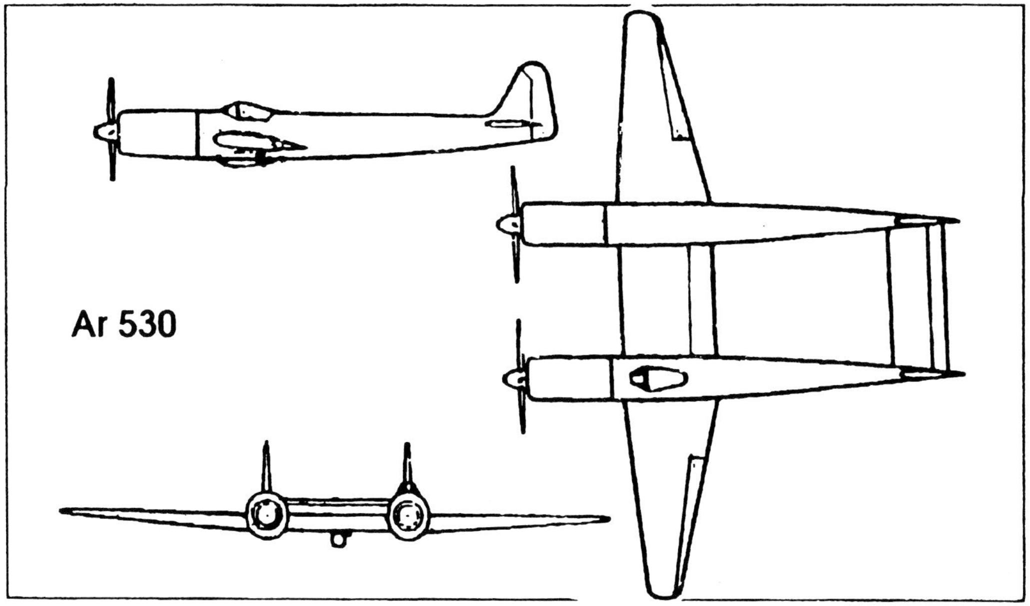

Arado Flugzeugwerke engineers were considering an asymmetrical planform for their proposed *Arado Ar P.530*, which was intended to be a twin piston engine-powered high-altitude day fighter. The pilot's cockpit was located in the port fuselage.

The proposed asymmetrical 3-piston engine seaplane...the *Blohm & Voss Bv P. 111.* Scale model and photographed by *Reinhard Roeser*.

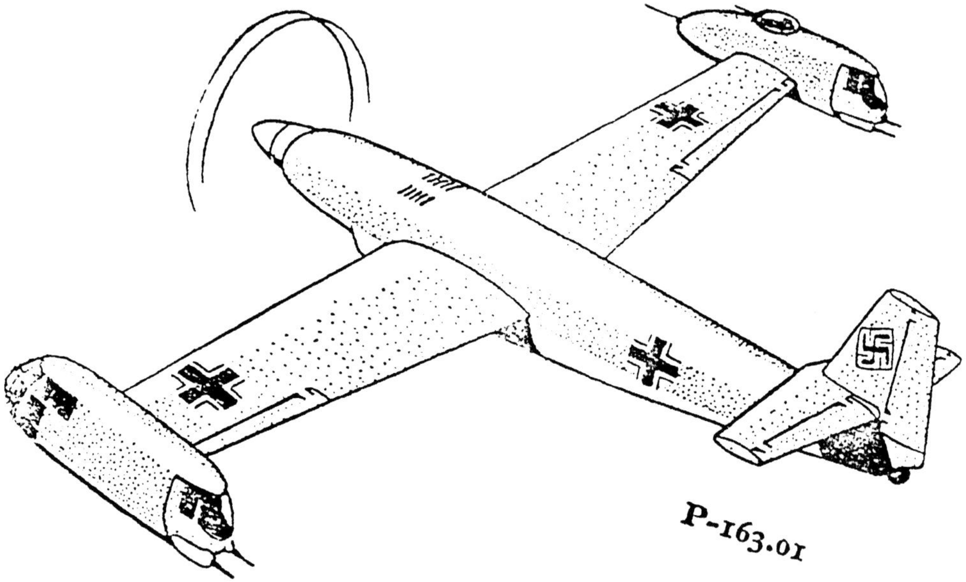

The proposed asymmetrical *Blohm & Voss Bv P. 163.01* general purpose aircraft. It was to have been fitted with a single *Daimler-Benz DB 613* 24-cylinder engine. Pen and ink drawing by *Hugh W. Cowin*.

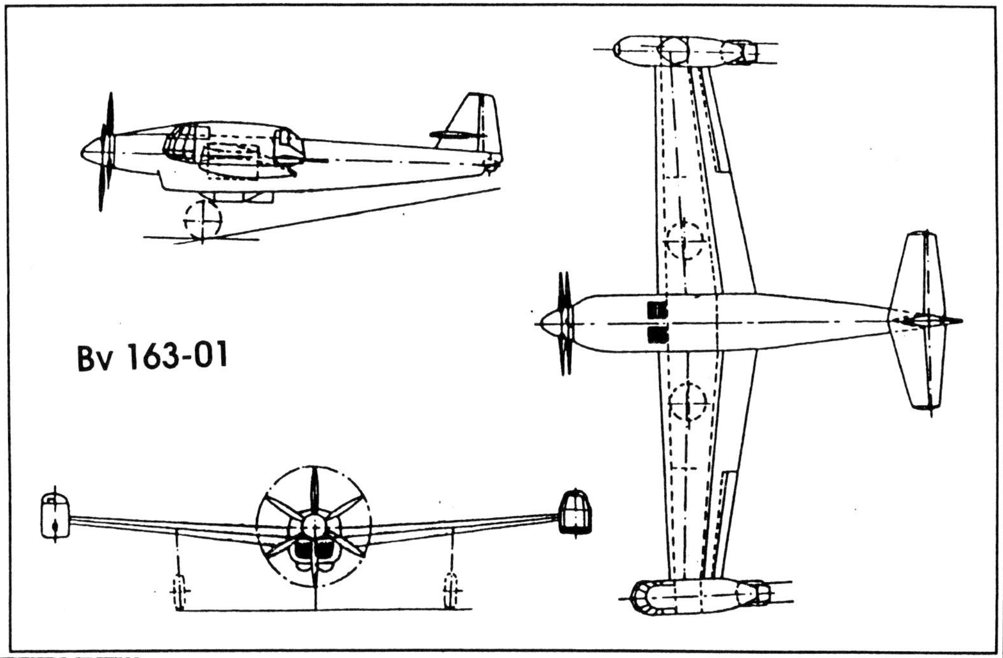

A 3-view drawing of the proposed asymmetrical *Blohm & Voss Bv P. 163.01* general purpose aircraft with its single *Daimler-Benz DB 613* 24-cylinder engine driving a contra-rotating propeller.

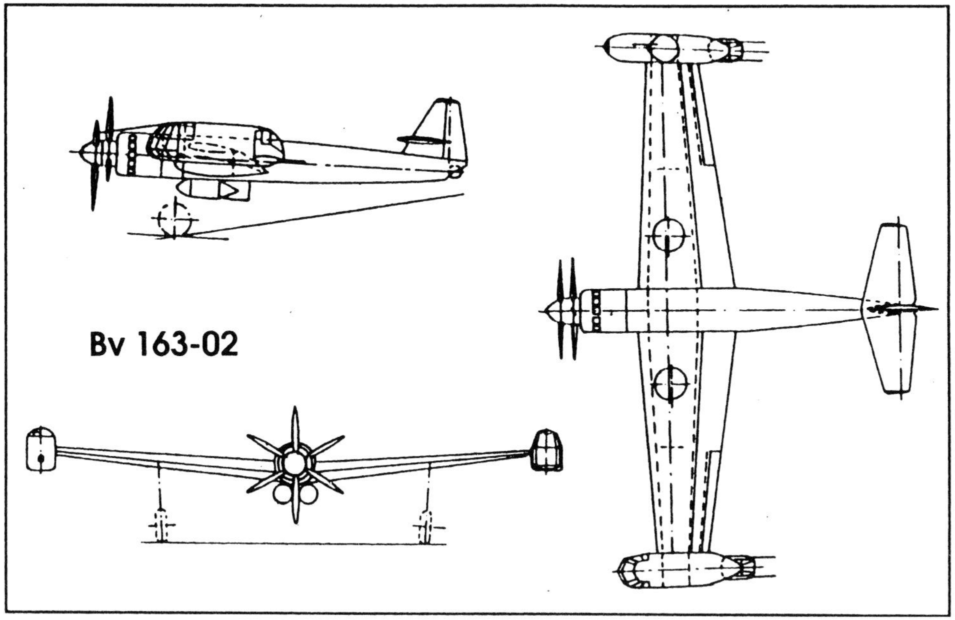

A 3-view drawing of the proposed asymmetrical *Blohm & Voss Bv P. 163.02* general purpose aircraft with its single *BMW 803A* 28-cylinder engine driving a contra-rotating propeller.

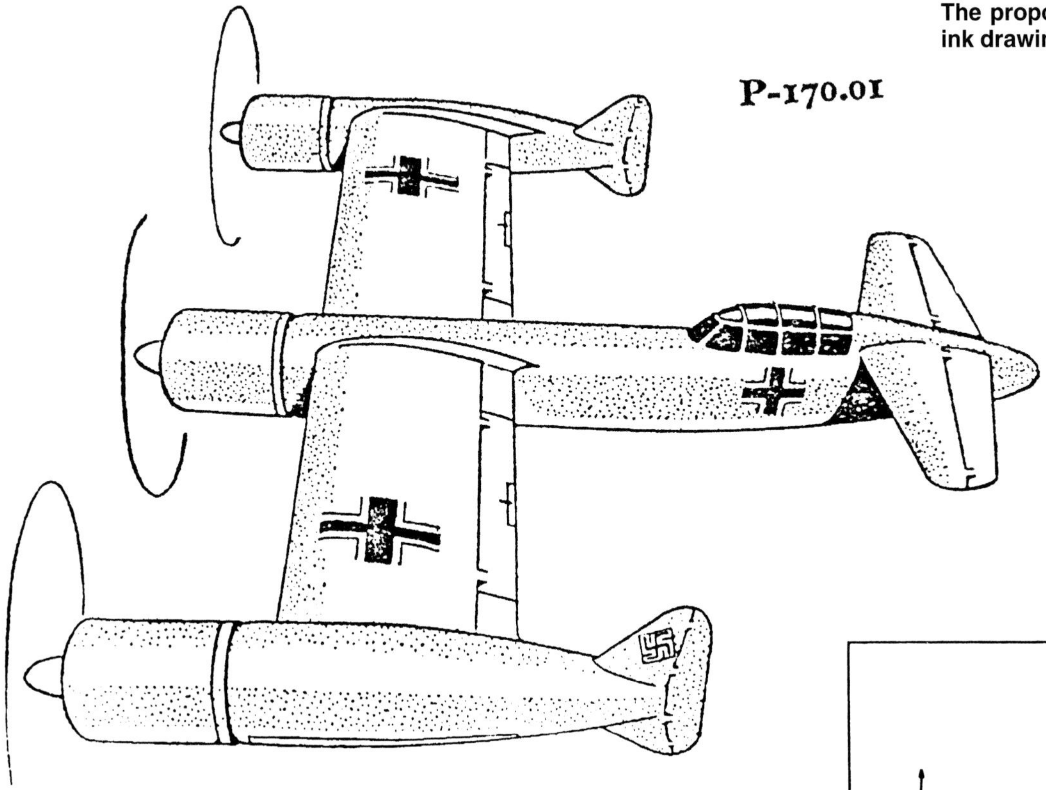

The proposed *Blohm & Voss Bv P. 170.01.* A pen and ink drawing by *Hugh W. Cowin.*

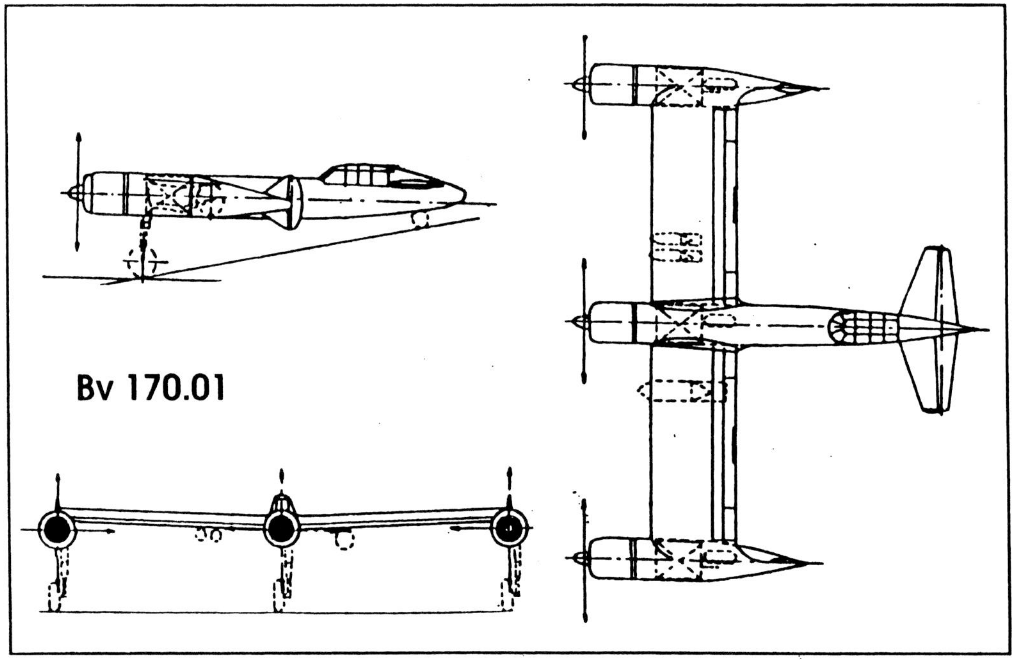

A 3-view drawing of the proposed asymmetrical *Blohm & Voss Bv P. 170.01.*

The proposed *Blohm & Voss Bv P. 170.01.* Scale model and photographed by *Reinhard Roeser.*

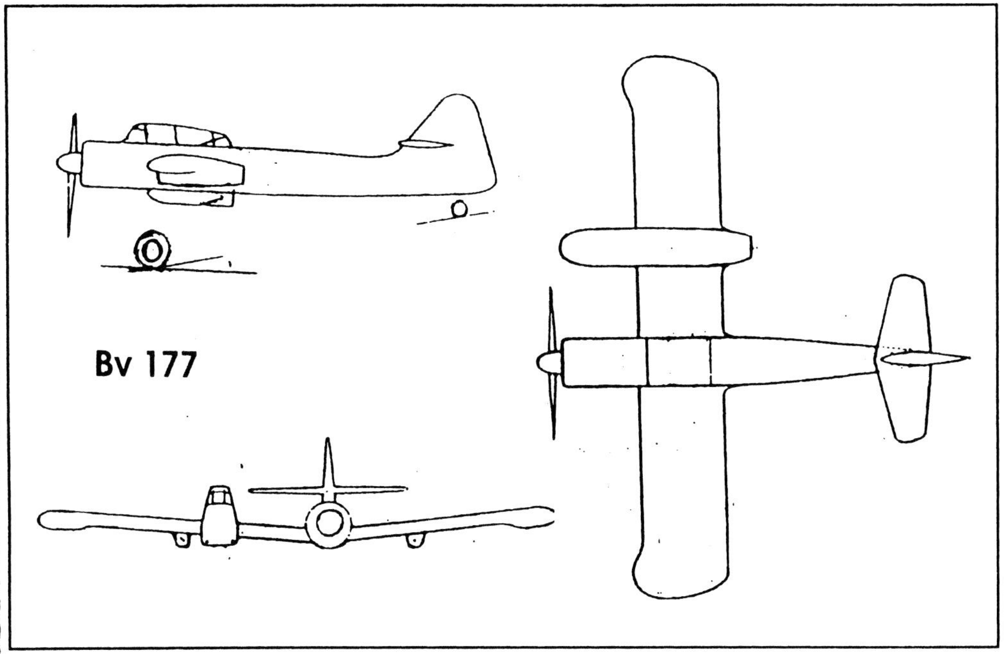

A 3-view drawing of the proposed asymmetrical *Blohm & Voss Bv P. 177* dive bomber and/or ground attack aircraft with a single *Junkers Jumo 213* radial piston engine located in the fuselage nose

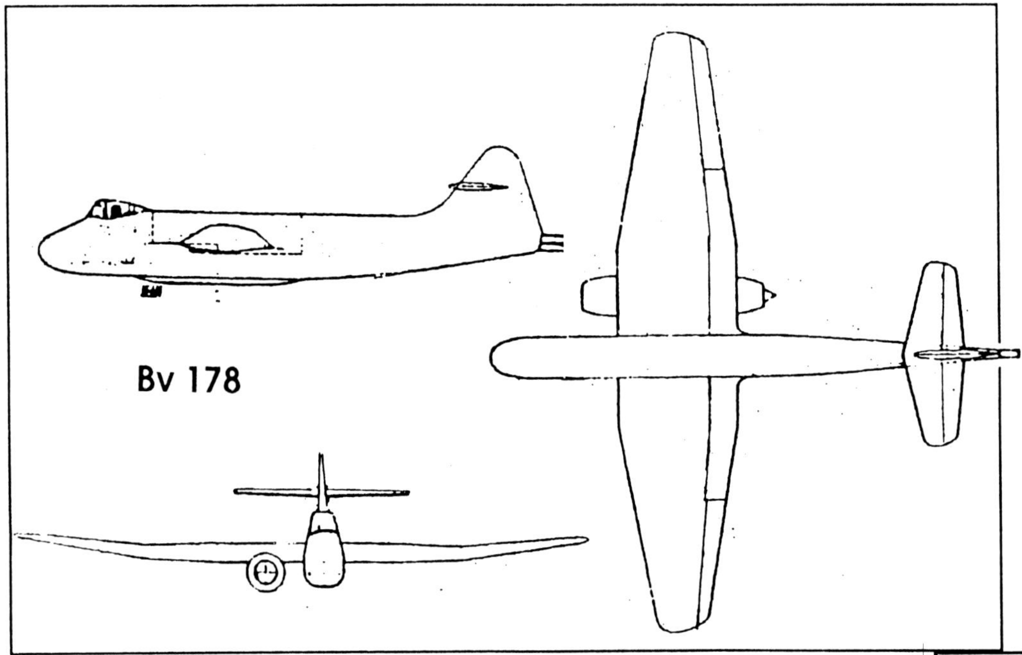

A 3-view drawing of the proposed asymmetrical *Blohm & Voss Bv P. 178* dive bomber and/or ground attack aircraft and powered by a single *Junkers Jumo 004* turbojet engine slung under its starboard wing.

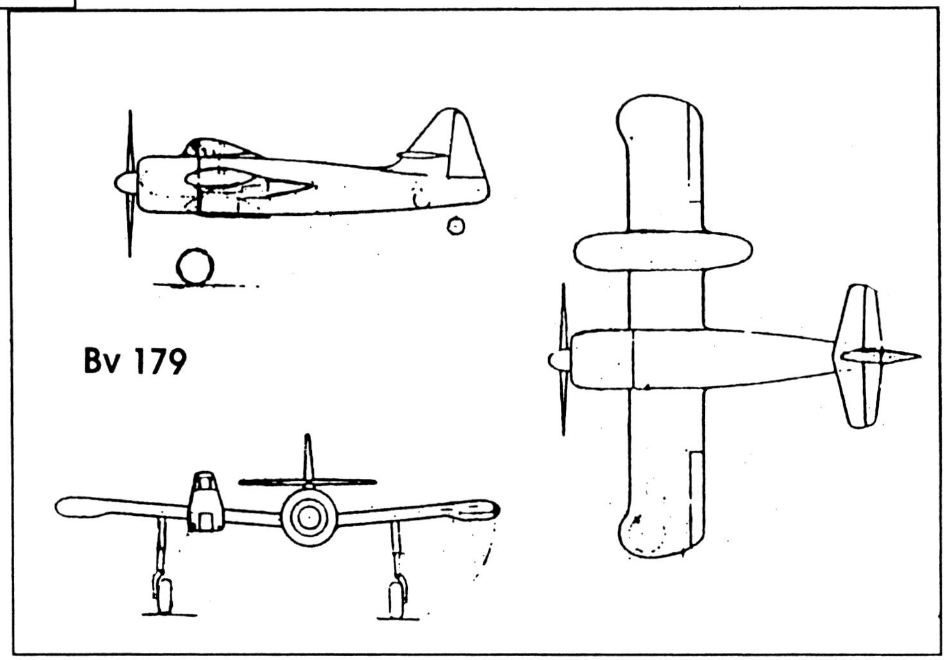

A 3-view drawing of the proposed asymmetrical *Blohm & Voss Bv P. 179* dive bomber and powered by a single *BMW 801D* radial piston engine located in the fuselage nose.

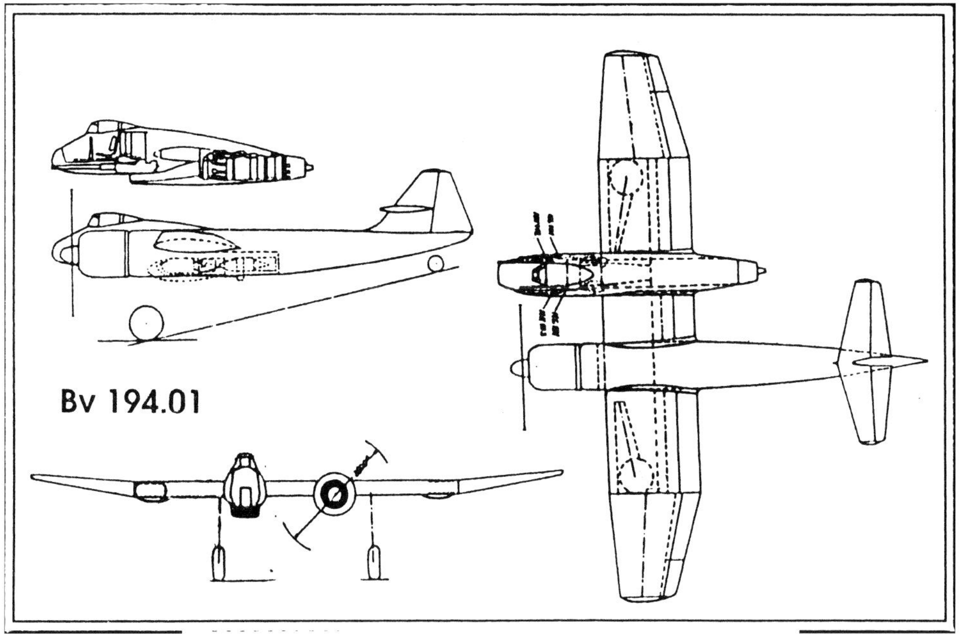

A 3-view drawing of the proposed asymmetrical *Blohm & Voss Bv P. 194.01* multi-purpose aircraft intended for ground attack duties. It was to have been powered by a single *BMW 801D* 14-cylinder radial piston engine located in the nose of the fuselage and a single *Junkers Jumo 004C* turbojet engine located within the cockpit nacelle.

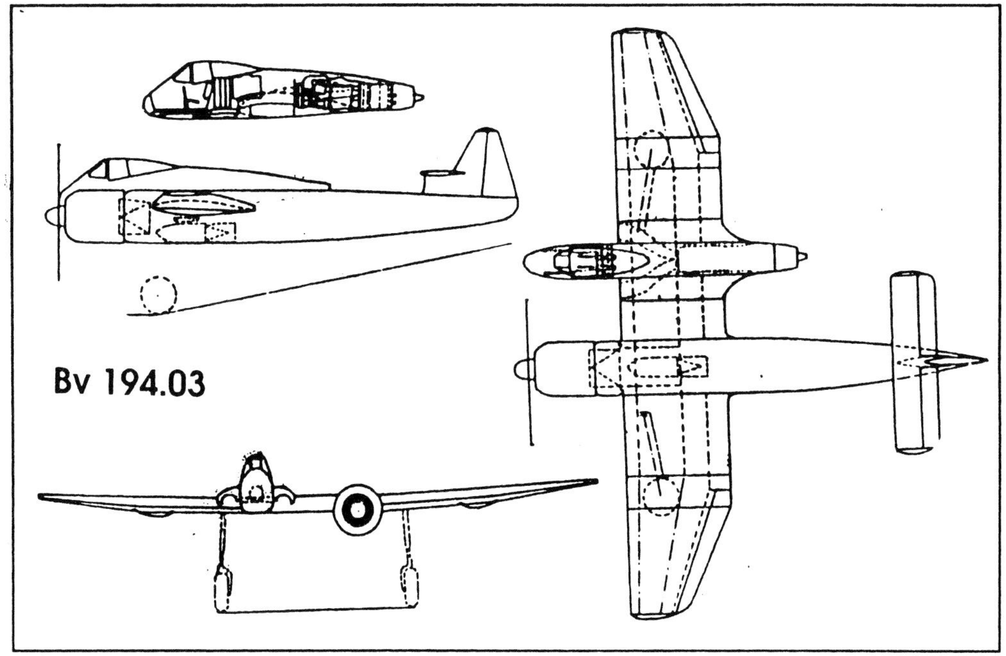

A 3-view drawing of the proposed asymmetrical *Blohm & Voss Bv P. 194.03* multi-purpose aircraft. It was to have been powered by a single *BMW 801D* 14-cylinder radial piston engine located in the fuselage nose and a single *Junkers Jumo 004C* turbojet engine located in the cockpit nacelle.

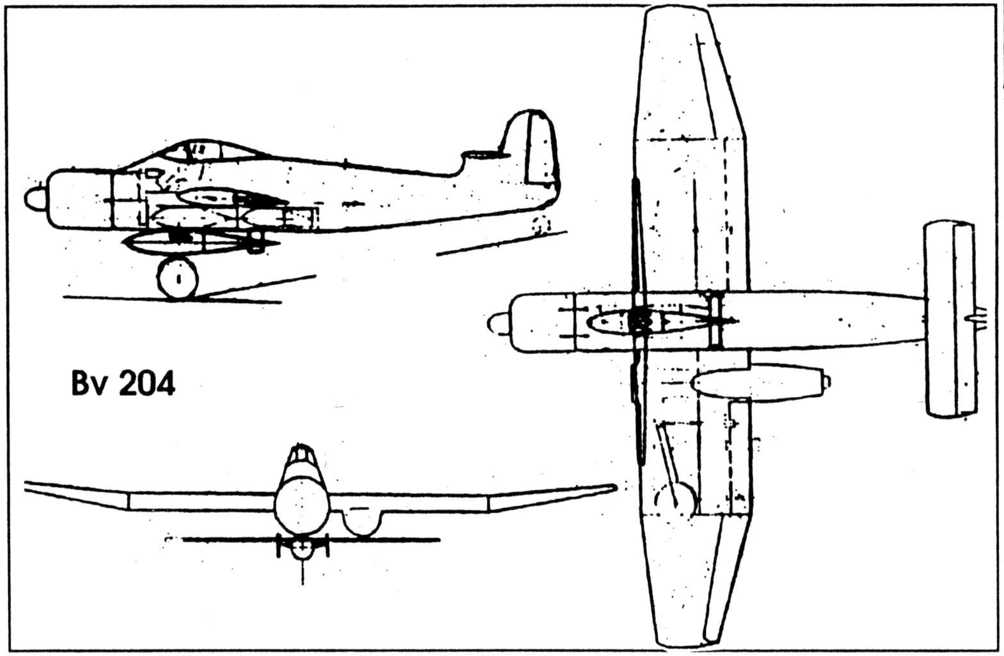

A 3-view drawing of the proposed asymmetrical *Blohm & Voss Bv P. 204* dive bomber and/or ground attack aircraft. It was to have been powered by a single *BMW 801D* 14-cylinder radial piston engine located in the fuselage nose and a single *BMW 003* turbojet engine located in a separate nacelle on the port side wing.

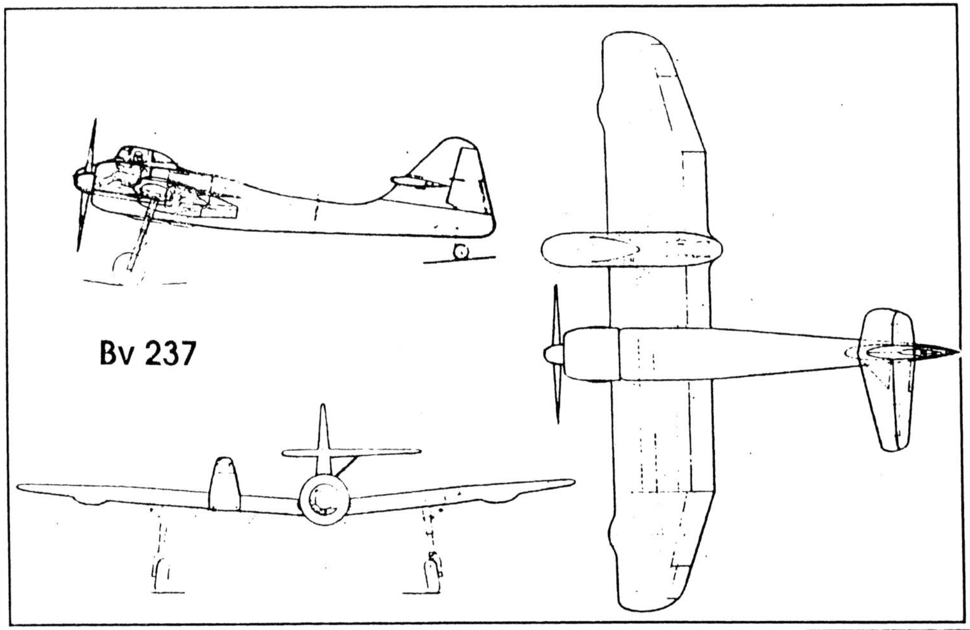

A 3-view drawing of the proposed asymmetrical *Blohm & Voss Bv P. 237* dive bomber and/or ground attack aircraft. It was to have been powered by a single *BMW 801D* 14-cylinder radial piston engine and a single *Junkers Jumo 004B* or *004C* turbojet engine located in the cockpit nacelle.

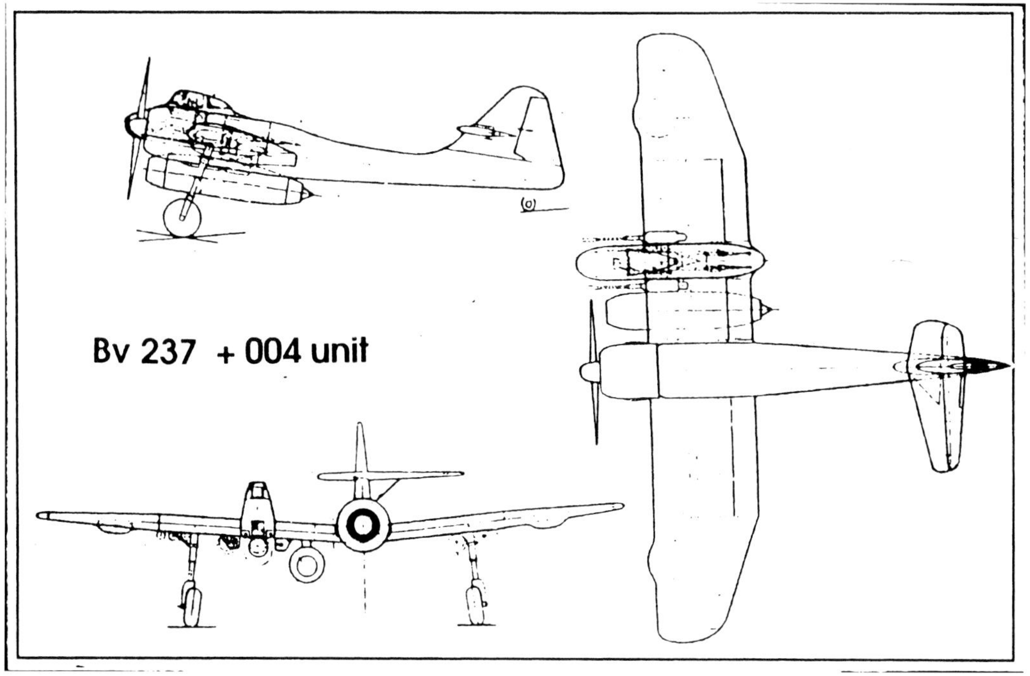

A 3-view drawing of the proposed asymmetrical *Blohm & Voss Bv P. 237* dive bomber and/or ground attack aircraft. It was to have been fitted with a single *BMW 801D* 14-cylinder radial piston engine and a single *Junkers Jumo 004C* turbojet engine mounted on the port wing's undersurface between the fuselage and cockpit nacelle.

An artist's illustration of the jet powered asymmetrical *Blohm & Voss Bv P. 237* dive bomber and/or ground attack aircraft and powered by a single *BMW 801D* 14-cylinder radial piston engine and a single *Junkers Jumo 004D* turbojet engine mounted on the port wing's undersurface between the fuselage and cockpit nacelle.

Digital image of the proposed *Blohm & Voss Bv P.170.01* by Kevin Wafer.

Digital image of the proposed *Blohm & Voss Bv P.170.01* by Kevin Wafer.

47